Do not just live, leave your trail...

I do not say this often or lightly, but you would be a very good Marine. We would be honoured to have you amongst us and I'm sure we will feel very comfortable within your ranks as well.

Lt Colonel Joe Collins, US Marine Corps, 8 Nov 2004.

...the loss is entirely of the organisation, to be denied opportunity of organisational growth possible by people of your calibre and humane approach and who are rare to come by.

All this ought to be untrue-your quality of head and heart are un-paralleled. We are too small in front of you.

Maj General Vinod, 11 Apr 2007.

Indeed, it was a pleasure serving with you. Seldom do I come across persons whom I admire in entirety. I must confess you are one of them. I appreciate all that you did for the Paltan.

Col Jacob Freeman, 9 Sep 2009.

On this last night in Lolab, I only think of you. We have never missed anyone so much. In fact, the 17 months spent under you are the most precious in entire service. The learnings, teachings and the essence of life acquired from you are going to stay with me always and every time. I assure you sir, that there will be no let-up in the motivation gained from the trust and confidence reposed by you.

Lt Col Vishal Chadda, 28 Dec 2009.

While ranks and achievements are important, it is other considerations such as being an 'evolved person' that count as much. Here you score an edge. Your journey reminds us all that we can make our lives sublime too.

Maj General Vinod, 26 Nov 2012.

... Your one tenure as the commandant of NDA will further add to the prestige and glory that NDA has achieved as one of the best training institute in the world.

Brig Ranbir Singh, AVSM, VrC, 14 Jul 2013.

With your positive attitude and large comfort zone, one has never felt the absence of guidance. We look up to you as an ideal leader and take pride in the same.

Col Amit Sood, 06 May 2013.

It is only our troops who can win our hearts and admiration with their dedication. This shows what you have done to the division you are commanding. Thank you for sparing time to meet me. I found it very difficult to take leave of you.

Lt Gen YS Tomar, PVSM (Retd), 23 Sep 2013.

Your style of command was really charismatic. Sir, if war breaks out then the Team Fourteen would be fortunate if we get another opportunity to serve under your able leadership.

Col Rohit Khanna, 6 Jan 2014.

It was a great honour for me and all ranks of 27 TIGERS to serve under a SOLDIER GENERAL like you. Your words of encouragement motivated us to strive for excellence in all fields. Your humble and simplistic approach made us feel confident and comfortable when you were around. On behalf of all officers of my Paltan, may I request you to accept honorary membership of our officers' mess to guide us in our future endeavours.

Col YD Mallur, 15 Jan 2014.

Interaction with you has motivated and fuelled me enough to carry on in this wonderful profession with same zeal and enthusiasm. I will forever cherish the part of my life time spent under your command.

Maj Pragnesh Pandey, 16 Jan 2015.

Courage requires having a voice—a voice that addresses difficult subjects, delivers tough feedback, and shares dissenting feedback.

This profession is unique and demands a unique approach like none other. Stand up and be counted, from day one.

FOUR DECADES IN OLIVE GREENS

Pride, Passion and Perspectives

You were and will always remain a role model.

Maj General Deepak Dhanda.

You can take immense pride in your long march from the day you passed out from the IMA to the day when you will finally bid Farewell to Arms. The likes of you have NEVER hesitated even for a moment to relegate your own self before the army and the Jawans. I dare say that all those who know Anil Sengar will continue to remember Maj General Sengar with great admiration.

Colonel Ashok S., Jan 2016.

Back Cover Transcript

Non-empanelment at some stage in life is a reality -even for larger than life officers like you. Towards this, I'd liketo tell you today (and a million times earlier), that you are a source of inspiration for a very large number of us. We in the battalion have looked up to you (almost revered you) as a role model. And it is not surprising that even after leaving the battalion, you have been a cause of cheer and happiness for a very large cross section of people. Putting it simply sir, you are one of the most successful officers I have known.

Your simple nature, no personal requirement attitude, professional brilliance, persona and commitment to the organisation are the traits that we have tried to emulate. You have left an indelible markon the battalion right from 2Lt days. The battalion and its team of officers has a major reflection of your contribution over the last three decades plus and you have been instrumental in bringing the battalion to where it is today.

Sir, regardless of the board results, I wanted to tell you that you remain the tallest officer/leader in my(our) eyes and we continue to draw inspiration from you. One man who will go home as a Maj General will be much happier than many others with an extra star.

Col Puneet Ahuja, 23 Aug 15.

FOUR DECADES IN OLIVE GREENS

Pride, Passion and Perspectives

Maj General Anil Sengar (Retd.)

PENTAGON PRESS

Four Decades in Olive Greens: Pride, Passion and Perspectives
By Maj General Anil Sengar (Retd.)

ISBN 978-93-86618-25-2

First Published in 2018

Published by
PENTAGON PRESS
206, Peacock Lane, ShahpurJat
New Delhi-110049
Phones: 011-64706243, 26491568
Telefax: 011-26490600
email: rajan@pentagonpress.in
website: www.pentagonpress.in

Printed at Aegean Offset Printers, Greater Noida, U.P.

To the Government

The willingness with which our young people are likely to serve in any war, no matter how justified, shall be directly proportional as to how they perceive the veterans of earlier wars were treated and appreciated by their nation.

—General George Washington

Contents

PART 2: MY PERSPECTIVE: WHAT AILS THE INDIAN ARMY?

THE BOOK IS NOT ABOUT HUMILITY

After four decades of service, I know that I earned myself the reputation of a man who walked the talk, whose word could be trusted and set standards of professional and personal conduct that many see worth emulating.

I feared no authority and no man; I only feared my conscience. I chased no ranks. I always chose the harder right, than the easier wrong. I never believed in apologising for doing the right thing.

I can foresee criticism coming on this book on many counts. No one kicks a dead dog and if you bother about criticism, you will attempt nothing worthwhile in life.

In a manner, I lived by the sword and fell by the sword. And I have no regrets.

For me, the measure of success lies in the positive impact one makes to the organisation and the people one is associated with.

Read this book and ask yourself this question, 'If you had the same potential and prospects, would you have had the courage to serve the way I did?'

ACKNOWLEDGEMENTS

I look at my life with great satisfaction. There is little that I would like to change. The credit starts with my parents and family. My wife handled the family single-handed with little or no contribution from me. She supported me very ably, enabling me to devote myself to my responsibilities as an officer without a worry. She sacrificed immensely for the children in every way. The entire credit of the good that exists in my family goes squarely to her. It is here that I would have liked to do better, if there was a chance. Arjun and Aryaman are wonderful boys who always did us proud.

My school, Ghorakhal, was instrumental in resurrecting me from a struggling kid to a confident or may be an overconfident adolescent who was not constrained by any boundaries or fears. The teachers of the time deserve full credit—so many of them.

My battalion played a major role in the journey of soldiering. The troops' simplicity, loyalty, sincerity and affection made me a better man and an officer. My first two commanding officers Lt Gen Pup Mann and Brig Umang Seth left a positive impact on my professional and social upbringing at a critical stage in my career. The unit officers, JCOs and NCOs, were excellent mentors who guided me in the early days.

My tenure as the Aide de Camp (ADC) with General Virender Awasty taught me values of honesty, simplicity, compassion, maturity and professionalism that stood me in great stead in my career. I have continued to enjoy the association with them since 1983.

Band of brothers in 8 Sector Rashtriya Rifles, 56 Infantry Division, 22 infantry Division made my journey an exhilarating experience. Their loyalty and commitment made those challenging assignments great fun.

In context of this book, I need to express my gratitude to General KH Singh, (Rajputs), a soldier of conscience, whom I have known from the National Defence Academy (NDA). He read the book and gave an honest and meaningful critique. His patience and personal knowledge were most useful in adding immense value to the book.

I need to thank my classmate from Ghorakhal, Mr Rakesh Pandey. He encouraged me to write this book, having seen me from the time we were ten. Without his push, it would have been difficult to cross the line.

I need to thank all those whom I was privileged to serve with. They all contributed in making me a better man, soldier and an officer. They made this journey a wonderful journey by their commitment and camaraderie.

A Tribute to My Parents

I am writing this four days after the demise of my mother. Eight months later, my father also passed away. As I write this, a major part of our lives seems to have been taken away. While they did not play an active role in our lives in the last few years, it is only after those beds have gone empty that we realise the monumental loss. I need to share their role in my life.

My parents were persons of meagre financial resources, my father having retired as Honourary Capt from the Army Education Corps (AEC) in 1983, four years after I was commissioned. But they were immensely wealthy in character, integrity, culture, values and love.

Colonel Sunil Attri, my coursemate, wrote on the demise of my mother that she must have been proud of me. I know she was. He wrote this because he has always appreciated the values that I have lived by and demonstrated over four decades. I inherited these values from my parents. My mother first prayed for the well-being of all other children and then for her own.

At the time, when I was selected for Sainik School Ghorakhal in 1968, my elder brother, Ashok Sengar, a well-known and respected paratrooper of 9 Para SF, was in school too. He realised that it was hard for my father to finance two of us in the boarding. He asked my father to send me and pull him out. My father didn't allow that to happen. My mother willingly sold her jewellery to send me to the school. And I learnt of this just a few months before she passed away at the age of 82.

After the demise of her mother at about twelve, it had fallen upon her to run the house with two younger sisters. She was a person of great inner strength and good values. She roughed it out as an Army wife and raised four children who inherited good values, had successful lives and earned good reputation in their respective fields.

My father's life was described by his contemporaries as the life of a saint. He was a highly educated and learned man; double MA and MEd. in the late sixties, having earned these qualifications while in service. He had to give up his pursuit of PhD due to the exigencies of service. He never drank or smoked ever. I never heard a word of profanity from him. He served with courage and stood by the military values as the Subedar Major of the AEC Training Centre, Pachmarhi. Post-retirement, he studied hard and taught students up to graduate level as a social cause, charging token money of Rs 90 per month when others were charging Rs 200 per hour, and teaching free those who could not afford even the token money. He spent a lot of time and effort studying health and yoga. He would photocopy thousands of pages of useful information on health and distribute them to the needy for free. He taught yoga as a social cause, much before Baba Ramdev made it popular.

He was bedridden in the last two years. In the period after the death of my mother, silently he taught us many lessons in courage, fortitude, compassion, courtesy and how to brave a huge personal loss.

A man of outstanding personal qualities, he earned himself a good reputation as a man of vision, values and compassion. This was reflected in a hand-written parchment presented to him by the Junior Commissioned Officers (JCOs) and Non Commissioned Officers (NCOs) of the AEC Centre on his retirement. Some excerpts are reproduced below:

> *"Apne, apni abhutpoorva karmathta, kartavya nishta, karyakushalta, drid vishwas, vyavharikta va doordarshita ka parichay dekar khyati hi nahi arjit ki, apitu sainik ke shiksha ke kendra mein anupam sahayata karke unke jeewan mein padonati hetu path prashasta bhi kiya.*
>
> *Aapka jeewan sant ka jeewan hai. Aapka nirnay swayam mein drid evam kalyankari hai. Parishram, lagan, spashtwadita adi, aapke gunon ne vibhin isham parishthithion men bhi apko safalta se ranchmatra door hone nahi diya. Apne shayogiyon evam shayakon ke sath apka priya, sammanyukt wa mitrawat vyawahar apki mahanata ka ghotak raha hai.*
>
> *Army Education Corps Centre ke vikasonmukhi karyon ko sajone men apka pathpradarshan ham sabko usi path par agrasit hone ki prerena dekar ja rahe hain.*
>
> *Anya ranks ke sabhi wargon ke sainikon mein jagriti lane hetu protsahan pradan karne ki adharsila apke kar kamlon se hi rakhi gayi.*

Apki yeh kriti sadaiwa is prashikshan kendra mein apki smriti dilati rahegi.

Apka yatarthonmukh adarsh hi apki uttarotra unnati ka mool mantra hai. Jahan ek ore aapne apna sainik jeewan sdhar banaya, wahin dusri ore apne pariwarak jeewan ko sawara hai.Apka parivarik vikas hum sabhi ke liye ek anupam udharan evam adarsh hai.

Ant mein hum apki abhivanjana ka rasasvadan, surdas ki in kariyan dwara vyakt karte hain:

JYON GUNGE MEETHE FAL KO RAS ANTARGAT HI BHAVE!
PARAM SWAD SAB HI SUNIRENTAR AMIT TOB UPJAWE!!

Together they made immense sacrifices to give us a good future based on good values and contributed unconditionally to the Army and the society in their own humble, yet selfless ways.

We mourn the void left by their demise, but we celebrate their lives for the magnificent persons they were.

'*Thank God for the parents you gave us. We wouldn't be what we are but for them.*'

My Biggest Regret

I left to attend the course get-together in the NDA when my mother was in an ICU in the hospital. Even though both my brothers were there to look after her, it was the biggest mistake, for which I cannot forgive myself. I returned after three days and never saw her again. I was always a good son, but faltered at the most critical time. The thought never goes away and it is painful.

When your parents leave, it is not their death that hurts you, for that is the eternal truth. What hurts you is, did they depart peacefully and did you do the best that you were capable of to look after them in their twilight years when they could not look after themselves? There is only one opportunity to look after them. Do not miss it.

As soldiers, we often hold back our tears and put up a brave face in the face of tragedy, personal or military, for crying is seen as unmanly for a soldier. It is okay to grieve for your dear ones and your brothers-in-arms. There is no shame in shedding tears for those who were part of your lives and made a difference to your lives. If you do not wish others to see your tears, stand under the shower and let them run down with the water.

FOREWORD

The book *Four Decades in Olive Greens: Pride, Passion and Perspectives*, is the third book by Maj General Anil Sengar; each book deals with a varied subject of immense interest to soldiers. He interacted with me during the writing of this book; as such, my views on important issues have been cited by him.

As the Commandant Infantry School, Mhow, I saw the interest his second book, *Battalion Command: Dare to Lead*, generated amongst the officers. Some described it as the best book they had read on junior leadership.

This book is written with passion and commitment, in an easy to read style with his characteristic humour. It is a reflection straight from the heart of a soldier, for whom the Army was not a career but a passion. He has aptly identified three areas that need deliberation if the Army has to sharpen the cutting edge of the sword, relating to the promotion policy, professionalism and an honest reflection on the senior hierarchy of the Indian Army.

I have handled these matters in different capacities in the Military Secretary (MS) Branch and others, and thus, have dealt with the subjects that the author has incisively dealt with. I agree with General Sengar that the promotion policy for star-rank officers, their postings and exposures to various responsibilities need greater attention. Discontentment amongst various arms and services due to the allocation of vacancies at the Colonel level needs to be looked at dispassionately, void of arm parochialism. I am sure that the issues that he has touched will help the Army to sharpen its sword and must be looked at by the hierarchy of the day. At the end of the day, it is ethical leadership which can provide direction to the soldiers.

General Sengar stands out as an officer who walked the talk and stood by his values irrespective of the consequences and is seen as a role model by many. His career has been a reflection of the values that he practiced through personal example. While the book may sound a bit narcissist, General Sengar has indeed earned his spurs well. The positive impact on those who were associated with him and the legacy he left wherever he served is indeed praiseworthy.

I compliment General Sengar, both for his courage in taking the bull by the horns by picking up sensitive yet critical issues, and for realising his dream of writing an absorbing book around himself and the issues he valued from his early childhood.

I have no doubts the young generation, now and in the future, will find the book inspirational in enjoying the grand journey of soldiering.

General Patton once remarked that if everyone is thinking alike, then somebody isn't thinking. General Sengar chose to think differently and write. I strongly recommend the book to be read by the officers at all levels.

Lt Gen KH Singh (Retd)

Former Commandant, The Infantry School, Mhow
and Corps Commander

Preface

I was fortunate to be in a profession I loved, was passionate about and wanted to be in. I loved the men that I was privileged to serve with in my battalion and other command assignments and I loved officers who formed my team. I loved the challenges that came my way as part of this profession, and there have been many. My almost four decades of service seemed to have whizzed past in a flash; this, for one simple reason that I loved every aspect of what this profession stood for. For me the uniform was not a three-metre Olive Green cloth, cut and stitched into a pant and a shirt. For me it stood for military values, tradition, selfless service, camaraderie, pride, moral courage et al.

I have been occasionally described as an idealist in a non-ideal world. I see that as an honest recognition of my efforts to do the right. For me doing anything else was accepting a compromise.

> *Ideals are like stars, you may never get there, but like the mariner, it helps you to keep direction.*
>
> —*Anonymous*

In the evening of their lives, more people will regret not living the way they would have liked to, for one or the other reason—mostly a fear of something, and they do not even know of what. I have no such regrets. I looked every superior commander in the eye. I chose to live the way I believed was the right way. That was never easy, for, few superiors appreciate a subordinate who looks them straight in the eyes.

My colleagues share interesting views about me. One is that I deserved a lot more in the Army. The other, considering my living by the sword, is about a surprise that I reached the rank I did.

My career is a case study with all the melodrama thrown into it. I had one of the finest records of service and a good reputation to go with it, all through my service. I had perfect nine reports; I also had a report with fourteen sixes in the cut-off report for promotion to Lt Col. My first three adverse remarks in the report came in the first three ranks of 2Lt, Captain and Major by formation commanders. I filed six statutory and non-statutory complaints; all were rejected. I was non-empanelled as a Colonel, yet I surpassed many who had all their tickets punched. It throws poor light on our complaint redressal mechanism, yet shows the system in good light, when manned by the right people.

Lt General Sangra, the corps commander, who was my senior reviewing officer (SRO), wrote in my cut-off report for the promotion board from Major to Lt Colonel:

> *"This officer will prove dysfunctional to the organization if promoted to higher ranks."*

Isn't that interesting, a corps commander writing this on a Major, whom he had seen for less than thirty seconds, and he is not even required to justify?

While my statutory complaint against this cut-off report was rejected, I did not miss the first two boards despite this report. To the credit of our system, while this report resulted in my non-nomination to the Higher Command Course, I was selected from ten high profile officers, mostly Higher Command Course qualified, for the post of defence attaché (DA), Turkey. So much for my being dysfunctional if promoted.

Years later, one Lt General met me. He seemed to know all about me. When I asked him, how? He said, he had handled my statutory complaint in the Complaints and Advisory Board (CAB). He said, while they were convinced that injustice had been done to me, no one had the moral courage to recommend against the SRO, who then was the army commander. So much for the high calibre officers of all ranks posted in the CAB and MS Branch.

For me, there was one thing about achieving anything in my life and career, that it must pass the ethical audit of 'the end, the ways and the means'.

It is easy to earn a reputation in battle if you are a good leader. It happens in quick time and operational achievements easily conceal character deficits. It is difficult to do it in peace and sustain it over four decades. I was part of many conflict situations that have existed in the last four decades. Introspection left me with a realisation that I had earned myself a universal reputation of a man of 'trust and credibility' whom many would like to emulate. Isn't that the crux of military leadership?

It is possible to serve with passion for everything that this profession stands for, selflessly and fearlessly in keeping with military values, enjoy the journey of soldiering and be a successful officer. But, to do that, you will have to rise above the ordinary and find a purpose that goes beyond you.

When I was superseded as a Colonel, one officer spoke to me and commented.

"You have been superseded despite your excellent record, but amazingly, you seem a very contented man!"

I said, "Yes indeed. Promotion to Brigadier, would have given me another two years of life of pride and commitment. I see that as an opportunity to contribute to the organisation on a bigger stage. I would have loved it. But, they cannot take away my fixed deposit of twenty-eight years of glorious soldiering, which I enjoyed every single day. I joined the Army for the love of the uniform, and not ranks. This is my source of motivation and I still wear that uniform. That is the source of my contentment."

"And what do you want to do now?" he asked.

"And now, my ambition is to be the best superseded officer the army has ever had."

I meant it and lived and worked that way.

Well, you can deny me a rank, but you cannot take away my passion. You can deny me a future, but you cannot take away my memorable past.

For one to write a book dealing with his experience, he needs to believe that they are worth a read. In my service, I have seen many senior commanders. In perspective, my principled path that never wavered, no matter the consequences, was unique and earned me many admirers and followers in the Army.

I do believe that my experiences are interesting and that many would wonder, are they really true? Yes, they are true.

This book is about my experiences as a soldier and the choices I made as an officer in dealing with various situations. These choices, never easy, were always made for the good of the organisation and at my own peril-never for my personal interest. I have no regrets for the path I chose. I wouldn't be writing this book if I was just another man looking for the next rank.

I am critical of four important issues—the promotion policy which is promoting mediocrity at the expense of merit, professionalism that needs an urgent upgrade at all levels, the quality of senior hierarchy and awards policy and practice. The most critical issues are the reforms in the Ministry of Defence and in defence procurement. They need to have the expertise and exposure to appreciate the complex dynamics of military requirements that cannot be understood by floating generalists. Equally important, there is a need to make them more integrated, responsive and accountable. However, this aspect is outside the purview of this book.

The most serious shortcoming in the Army is that an average and unsuitable officer can rise to the highest rank, by being 'system fit', a term I have talked about in the book. Competence and character is not a necessary condition for promotion in today's environment, as Admiral Arun Prakash, former Naval Chief, wrote in an article, "and it should not come as a surprise, if a number of undeserving "bad eggs" slip into higher (and highest) ranks." Surely, this is fatal for the military.

It is purely with the noble intention to redress the shortcomings of the institution that I served and loved no less than anyone else. The first step to do so is to acknowledge that problems exist, something our hierarchy is shy to acknowledge and act on.

It is not that I have suddenly acquired courage post-retirement; I had raised the same issues in all forums whilst in service including to General Bikram Singh, the Chief at the time, as the chairman of a study and possibly paid for it.

My criticism in no way takes any credit away from what the Army stands for despite the challenges. It has and will always deliver despite its shortcomings. It has set such high standards which other institutions can only dream about. It is the mainstay of the country. The Dera Sacha Sauda incident in Haryana in August 17, where a small state with a number of overpaid DGPs and IAS officers stood clueless and waited for the Army led by young Majors to restore order and enter the Dera Ashrams is a testimony to the Army's credibility and indispensability. The country sleeps in peace, only because a soldier stands guard on the frontiers, when others are either counting their votes or their dollars. My criticism comes out of my belief that all components of the Army must rise to the highest standards of the Army. For us there are no half-measures in any domain. That is our creed.

Where needed, I have sacrificed humility to convey that I am giving no moral lecture, that I practiced what I am preaching. A book whose basis is principled leadership would lack conviction if I criticised without stating what I did in those circumstances.

Take it or leave it; even as a student in school, I believed that one day I will write a book around myself. Here I am.

You are not bragging if you have done it

My Memorable Moments

Hav Bhupal Singh, 5 GUARDS

You just have to nod; 99 percent of the battalion will die for you (I was sad, for there were still one percent of men whose trust I could not earn!)

*

As Commander, 8 Sector Rashtriya Rifles (RR), Lolab

From a Commanding Officer, 'the men believe that you have brought success and good luck (Not one man suffered a scratch in combat while eliminating large number of terrorists).

*

As the General Officer Commanding 22 Infantry Division

From an MT Driver

The fuel expenses of the division have reduced by half, since my arrival.

*

A Tribute from Officers

- While the officers will miss you for sure, the men of the division will miss you even more.
- There are many Generals who are claiming the tag of a 'Soldiers' General': You are our concept of 'Soldiers' General.'
- If there is an opportunity to fight a war, you are the one we would like to fight the war under.

  And finally, the one I am most proud of:
- When my baby is born, I would like General Sengar to give him *ghutti* (honey). In our community, it is believed that the baby imbibes the character of the person who gives him *ghutti*. (A young expectant mother in the division)

This makes me hang my boots with the pride that I earned my spurs well as a soldier.

PART 1
REMINISCENCES

GROWING UP IN GHORAKHAL: THE GIFT OF A GOOD START

Dear Anil,

It was really a pleasure to go through your mail. It felt as if I am conversing with you in person across the table.

Your narrative to sum up your military life was very impressive and interesting, befitting your persona. I can empathise with it, being a knowledgeable friend and a classmate from those seven years in Ghorakhal, when we grew up together with our personalities in transition.

It was here we could have a glimpse of the soldier and the army officer you were to become; always bold and daring, a leader of the people who could take everyone along and also took a stand with authorities. Despite being the school vice-captain, you were more popular than the best cadet due to being multitalented and full of energy. Thus 'the General Sengar' was apparent in actions and deeds/misdeeds during the school days. For us it is an honour being a classmate of yours and being able to say that we know you as a friend and a classmate.

Our classmates are especially proud of you. I consider you to be a successful person who has charted his own path, as God has blessed you with some divinity, in the sense that you realised your potential from early childhood.

Now, this also deserves a biography from you to be penned for inspiring like-minded young minds who wish to chart out a similar course in life.

Rakesh Pandey, SSGK Roll No 348

(Classmate from school.)

*

Suhana Safar aur yeh Mausam Haseen. This famous song from the famous film of the sixties *Madhumati*, which was shot in the locale of my school, sums up our glorious journey in the school. The school is a paradise on all counts and we were blessed to have studied there.

Incidentally, my glorious journey in school started on an unpleasant note, which in a manner, set the tone for the future.

It was just a few months after I had joined the School, in March 1968, all of ten years old, that I found myself standing outside the Principal's office, like a convict. There was a flurry of activity as the Head Master, Flt Lt TB Srivastava and my House Matron Mrs Rovina Ray walked in and out of the Principal's office. I was the cause of the commotion.

The Head Master led me to the Principal's office and sat down on a chair on one side. In an oversized blazer that was meant to last seven years, I stood unfazed, facing a very pensive Principal.

"I am shocked that a fellow of your age, who has hardly stepped into the school has the cheeks to slap his matron?" he said without any formalities.

"Do you have anything to say?"

I replied in broken English, "This is a cold place, sir. I feel cold and the matron does not allow us to use the blanket issued from the school. I decided to use the blanket, against her orders. She got to know and grabbed my ears and yelled at me in front of the whole house. I pushed her hand from my ears, but somehow, my hand got her face and the push turned into a slap."

"By this evening you will leave the Holding House (meant for new arrivals of class 5) and move to Singh House. Head Master, please ensure," The Principal directed.

By evening I was in Singh House, the lone recruit in no man's land. There was no dearth of volunteers to help me with my box and 'hold all' over scores of stairs in that undulating terrain. I was an instant hero in the school, a young fellow who had the courage to challenge an unjust order. That, I realised soon, was part of my DNA. My stay in the school had begun with a bang of an unpleasant kind. As time passed, I inspired a lot of students throughout the course of my stay in school for good reasons and my stories survived there long after I had left the school.

That trait remained with me, and in a manner, while making me the straightforward conscientious man that I was, also put me in a confrontation mode with incompetent superiors. A compromise with values was simply something I was not prepared to do, no matter the consequences.

Ghorakhal is located in God's own country, the Dev Bhoomi, fifteen kilometres from Nainital in Uttarakhand. It really is a dream location for a military school, located well over six thousand feet with perfect climate.

A serene quiet place, we grew up in the lap of nature; amidst the sounds of whistling pines, towering deodars, chirping birds and the chiming bells of the Golu Devta mandir.

The parade ground overlooks the Bhimtal Valley that contains numerous lakes, surrounded by beautiful hills. On the other sides, lush green mountains of pines rise into the skyline. Beautiful deodars and the blue sky complete the majestic panorama. The famous Golu Devta temple overlooks the school in divine grace. He stands sentinel and his blessings are spread to all corners of the school by the cool breeze rustling through the pines. The chiming of the bells in that serene environment truly takes one as close to God, as one ever can.

The road ends at the school main building, which means there are no casual visitors. When I was a cadet there over forty years ago, the only connection with the rest of the world was a solitary UP Roadways bus that came to the school once a day. Forty years on, nothing has changed, except that the bus has become a Uttarakhand Roadways bus. The cadets grow up in that beautiful Shangri La, the very lap of nature, far from the crooked world. It does not get better.

Ghorakhal is an unusual name. The school is located in the old palace of the Raja of Ram Nagar. The legend has it that during the first war of independence in 1857, a British officer on the run from Ranikhet landed at this location. His horse, while drinking water from the twin ponds, now in the school campus, collapsed and died. Hence the name, Ghorakhal, meaning the horse pond.

Sainik Schools are the feeder schools to the National Defence Academy (NDA). Joining the NDA was the sole aim. You could become the Prime Minister, but if you did not go to the NDA, you were no good. Ghorakhal has credited itself honourably in this regard, lifting the Defence Minister's Trophy more often than any other school.

Life in the school was an absolute bliss. Those days, grades did not matter as long as you went into the NDA. The school had a variety of co-curricular activities which the modern schools attempt to provide today. There was no pressure of studies. I loved debating, dramatics and sports, was twice awarded the coveted best actor golden certificate, the school equivalent of the Oscar award! I loved public speaking and represented the school in inter-school debates in the country. Studies happened as a by-product. Most of our early years were spent cutting the hill sides and orchards to make sports fields. In the absence of sports infrastructure in the early years, we became natural runners, as that was the major sports activity in the school and came to be known as 'Ghoras', just the right name for the old boys of Ghorakhal.

A fun-loving character without a worry on earth, I enjoyed life at school, every bit. Everyday during the school assembly, one student from each class by turn was required to march up in front of the parade and speak something. A Class V cadet was required to deliver the 'thought of the day' and a Class XI cadet was to deliver the extempore speech on the given subject, etc. I was the ready substitute for Class V classmates who were too nervous to speak. One day, as the last moment substitute for one, I walked up without an idea of what I was going to say, for I knew I would figure out something. As I marched those forty yards to the mike, I decided and spoke:

> *"There is no life without wife." And as there was laughter all around, I finished adding, "And wife is a knife to cut the life." This remained in memory of the school for long.*

Everyone wondered where had the wisdom come to me so early in life from!

The teachers and the staff were superb. They knew their job, were very devoted and loved being part of the school. We can never thank them enough. The Ghoras are unanimous in their belief that we were the chosen few to be in Ghorakhal, and the school turned our lives around. Growing up in that remote but a heavenly place, the students grew up as honest, simple, hardworking and trustworthy individuals, typical of the highlanders.

I struggled in academics in the early years. On the first day, we were interviewed to screen the students who knew English and those who did not. I was asked three questions: What is your name? What does your father do? And, where do you come from? I answered with the confidence of an English graduate.

Section A, the teacher said, and I joined the English-speaking students. Over the next few days, I understood not a word of what was being taught in the class in English. I walked to the teacher and requested to be sent to Section B, which had students who were weak in English.

The teacher said, "I remember you replied to my questions very well, that is why you are in this section." I told him that these were the very questions I had prepared for the interview for selection to school and "you asked me the same". I was sent to Section B.

My brother, a senior in school, helped me immensely to improve by writing for me questions and answers for various subjects. He later joined 9 Para Commando and earned himself great credit as a soldier and a human being of substance.

The first few years were tough, as I struggled to survive academically. In the previous school, I had passed classes III and IV with grace marks and barely squeezed through with the teacher's mercy. Every odd day, I and a couple others were taken to the Principal for poor performance and not doing the homework. We looked forward to that as the student who took us to the principal was a pretty girl, whose brother landed up as a course junior in my squadron in the NDA ten years later. How God connects, one never knows.

My parents made immense sacrifices to put us in that school. My mother had to sell her little jewellery to raise money to send me to school, something she told me only last year. Ghorakhal was the best thing that happened to me. I realised very early, as my classmate mentioned, that I was gifted in many ways, and realised my potential early. I also realised early that I had to write my own script for my life and be the central character of that script, that I alone could make something out of myself and no one else. By the seventh class, I had caught up with the rest and was on my way and never looked back. The school was a boon.

Coming from a humble background, I realised the value of hard work, resources and money and used money with caution, never wasting one penny. Of the six rupees pocket money we got every month, I would save some and take it back home and gave it back to my mother. She would scold me as to why I did not spend it to eat and drink milk, etc. These two habits stayed with me forever. I never wasted or misappropriated any resource, whether it was my personal or the organisation's. In fact, I could never overlook an unattended running tap, even on the road side, and would turn it off.

The school was great fun. The teachers were dedicated and simple.

Those days in the seventies, there were very few magazines. Each house got one *Illustrated Weekly*. Khuswant Singh was the editor. I guess, going by his reputation, the magazine contained one juicy photograph, a model in a two-piece bikini. All that the cadets were interested in that magazine was that one photograph. Mr B.N. Srivastava, a brilliant mathematics teacher, was the house master of Kesari House. Stockily built with a broad chest, he had a typically broad nose, nasal accent and a conspicuous bouncy gait. He was genuinely concerned that the cadets will get spoiled by the photograph. So his master plan was to first flip through the magazine himself and pour over the photograph to absorb the details. He would then laboriously convert the two-piece bikini clad model in a swimming pool into a salwar kameez-clad model with a pen.

A simple man, he had a tough time managing the mischievous cadets as the house master of the Kesari House.

This one incident amply highlights the challenges the house masters faced on a daily basis.

The cadets' wash used to be sent together for each dormitory. Everyone had their school numbers written on their clothes. So a dormitory would send 30 odd sheets, 25 shorts, etc. They would be brought back after the wash by the school washer man, Munna Lal, as stacks of each item. The distribution drill was well organised. The wing prefect would take count of the clothing delivered. The cadets would line up in the dormitory, and he would call out the school number on the clothing and the owner would collect his item.

Once it happened that the wash was being distributed after dinner. There was one bed sheet left without any number on it and two cadets claiming it. Both showed their list of clothes sent for the wash to stake their claim on the lone sheet. Clearly, thirty sheets were sent for the wash and thirty received back. So some smart alec had whacked one. This led to immense commotion as the prefect was not able to resolve the crisis.

In walked the house master, Mr Srivastava. He spoke in his nasal voice and heavy eastern Uttar Pradesh accent.

"Yes prefect, *whaat* is the matter, *whaat* is this commotion about, *whaat* has happened?"

The prefect explained the warlike situation.

The house master, a simple honest man, was visibly disturbed that there was one dishonest man in his house.

He chose a novel way to resolve the issue. He first gave a long pep talk on morality, integrity and honesty. Mahatma Gandhi's experiences with truth were referred to and Gautama Buddha also cited generously.

Then, having laid out what he would have imagined an inspiring background, he appealed to the good sense of the cadets.

He said, "Thirty bed sheets were sent and thirty were received back. So, someone has taken an extra bed sheet. Remember, honesty is the best policy and this is not expected from you, the future officers of the Indian military. I will now switch off the light for one minute and the cadet who has taken the bed sheet should quietly keep it back and no one will know who he is."

The light was switched off for one minute. He then asked the prefect to switch on the light. When the light was switched on, the one remaining bed sheet had also disappeared without a trace.

Shakespeare would have found it difficult to describe the expression on the face of Mr Srivastava. He should have realised that in a military boarding school, good senses take much longer to develop!

There were icons that can never be forgotten. The school bus was a second-hand Ford on the edge of its grave. Through sheer will power it would manage the uphill drive to Nainital, but downhill, it could beat Mercedes. The driver, Nasrullah Khan, loved it. He took his own time driving, never in a hurry to reach the destination. As a result, half of our outing time to Nainital was spent in the bus. With no more than three rupees at hand for the outing, we did not mind.

He would often complaint to the school authorities that the school needed a better bus. One day, the Principal asked him, how was the bus?

He replied, "*Bahut accha hai saab, is mein horn ke ilawa sub kuch bajta hai.*" (It is great sir, everything in the bus except the horn makes noise).

Visibly embarrassed, the Principal asked him," "*Is me kya karne ki jarurat hain*?" (What is required to be done in the bus?)

Nasrullah replied, "*Jiada kuch nahi saab. Sirf do cheez badalne ki jaruruat hai. Ek toh engine, aur doosra body. Baki sab theek hai* (Nothing much, just need to change two things, the engine and the body).

As happens in a boarding school, pranks and mischief occupied us most of the time. Dense fog was common place in monsoon and winters. A clear day could turn to a white-out in minutes and you won't see your hands. The dense wet fog was depressing, which sometimes lasted for days. It was also an ally and the time to strike. The school was surrounded by lovely orchards and fruit gardens. We feasted on farm-fresh delicious fruits.

May be, on one occasion I tried my luck too far. I with my friends decided to raid the apple orchards of the Principal's house for a kick. As quickly as the fog had descended, it disappeared with the wind. His servant raised a frantic alarm as if the Principal's house was on fire. We took off chased by the Principal, Wing Commander Mukherjee, in his jeep. We tried to beat him by climbing on to the hills. Alas, we were recognised. Piled up in the principal's jeep, we were taken to the school main building and locked up in a garage. The school was made to fall in and we were lined up as criminals and given the dressing down of our lives.

In my last two years, my nonchalance and devil-may-care attitude did not find favour with the headmaster, Flight Lt GSK Yadav. I was doing well as a student; represented the school in debates in the country, played all games for the school, excelled in dramatics, was in top bracket in academics and was a kind of inspiration to many even at that stage. For sure, I loved pranks and was not too bothered about rules and regulations.

One day, I was called by him to his office.

"Where were you last night?"

"I was in my room, sir."

"Don't try to be smart, Sengar. I have straightened many like you. I know where you were last night. You are very close to being withdrawn. So you better tell me the truth, if you want to stay in the school."

"If I may, may I know where you think I was, sir?"

"I saw you enter the room of the matron in the Holding House. In that freezing night, I stood behind a bush the whole night, waiting for you to come out. But you never came out and I had to leave. You are simply lucky, I could not catch you red-handed."

I told him that he was mistaken. I know the matron well because of dramatics. She was our teacher like the rest and our relationship was that and that was it. The matron was an excellent caretaker for the young kids. She looked after the first year cadets, nine and ten years old with great care, concern and love. The kids kept walking in and out of her room with one or the other problem. This headmaster had let his imagination run wild and had frozen in that cold night. I guess he deserved it.

Disgusted at his failure to nail me, entirely on his wild imagination, he awarded me a two-hour pack drill along with a senior from Class XI for some other offence. The drill instructor, a Non Commissioned Officer (NCO) of the Army, took his order seriously as the headmaster stood and supervised our punishment himself for a while. About three quarters of an hour into the punishment, the other student collapsed exhausted. The drill NCO panicked and called the nursing assistant and asked me to stop. I said, "No way, I will complete the punishment." While the other cadet was evacuated to the infirmary, I stopped after two hours and went to my house, drenched in sweat, to a hero's welcome.

That was the maiden pack drill awarded and executed, as it was discontinued thereafter.

Much to the chagrin of the headmaster, I was appointed the school vice-captain. While many thought that I should have been the school captain, my fearless demeanour and equation with the headmaster played its part. The school captain VP Singh, was a very mature cadet and the right choice. There were many a clashes as our turbulent equation got no better. I was on his withdrawal warning list as the school vice-captain.

In the final year, we had qualified for the NDA and just waiting to leave the school. There are no words to describe that period of gay abandon. That success had to be celebrated. We raided the school baniya 'Shah ji's' shop at night. The intention was to whack sweets and distribute them to the cadets. We grabbed a couple of cartons in the pitch-dark night. Later when we opened our spoils, to our horror, it contained hundreds of Jai soap cakes. The school bathed with Jai soap for months.

Every day in school was fun. A classmate, Joginder, always up to some mischief, had his heart in everything except that mattered—studies. Son of a staff member, he squeezed through Class X with the grace of the Principal. On day one of Class XI, I saw him feverishly making notes. I thought the guy is turning over a new leaf and working hard. In the break, as I was walking past his desk, I noticed, he was making notes on small slips of paper about the quarter size of a notebook page.

I asked what was he doing, making notes on slips of paper?

He said, "Avoiding double effort yaar. Why first make notes in the notebook in the class and then before the exam, make '*purchies*'. I now take notes straight on '*purchies*'. Saves a lot of time and effort, you understand?"

I realised at some cost that one can get into trouble for no reason. There was one incident that happened in my final class that I cannot forget. My favourite subject was English. The teacher introduced the expression 'I couldn't care less'. The expression is used as such for all tenses. I was uncomfortable with this expression, as I felt it should be 'I wouldn't care less' for the future tense. So, I would often practice and speak this phrase aloud for no other reason than to get familiar with it.

My class was taken out for a two-day picnic to Kausani. One of the teachers on board the school bus was an unpopular character. I sat on the last row of the bus and the late Flight Lt Pradeep Rastogi, a brilliant student and an outstanding pilot, who died young in a crash attempting to save his trainee at the Air Force Training Academy, sat right behind this teacher in front. The class relentlessly taunted this teacher with his given name, which was no secret to anyone including the teacher himself. Myself and Rastogi were in the forefront of the assault as both of us had some issues with him. On the second day, as the bus slowly chugged along on the steep mountain road, spewing black smoke in its trail, this teacher suddenly turned around and started abusing and beating up Rastogi. That took the Principal, Lieutenant Commander GD Singh, also in the bus with his family, by shock surprise. He too got up and shouted at the teacher, asked him to mind his language, shut up and sit down.

The teacher, trembling with rage, shouted back, "Sir, these idiots, Sengar and Rastogi, have been taunting me ever since we left the school. I have tolerated this for the whole day and I can't stand it anymore."

The Principal ordered that he won't like to hear one word and everyone was to shut up till we reached the school. Fifteen minutes later, the bus halted for a short break. Everyone stepped out to stretch themselves. As was my habit, subconsciously I spoke out 'I couldn't care less', for no reason. The Principal, who was standing right behind, blew his top and shouted at me. "Come here, Sengar. You are the most bloody-minded fellow I have seen. You dare tell me, you couldn't care less after all that has happened less than fifteen minutes before? See me tomorrow in my office and you speak one more word, you are in real trouble." And I was the school vice-captain!

The next day I went to the Principal's office. He conveyed his immense displeasure over the previous day's incident. He particularly said that my comment 'I couldn't care less' in his presence was most horrible, for which he can never pardon me.

I told him the whole dynamics of my speaking out that phrase, accepting that it just happened at the wrong time and place.

The Principal said, "You expect me to believe that? Sorry I don't, you can leave my office."

What a quandary, my truth had no takers, but it remained with me. Almost six months later, we were finally leaving the school and on our way to the NDA in due course. On the last day, I went to his office and said that I wished to speak to him. He was surprised to see me since he had made his farewell speech to the whole class.

I brought up the old incident of 'I couldn't care less', during the picnic. Yet again, I explained with all earnestness the truth of why I spoke that phrase. I emphatically stressed that I was leaving the school that day and need not have broached this forgotten incident which was dead and buried. I only wanted the principal to know that I never meant to embarrass him, and what I was saying was the plain truth and would like to part with no misunderstanding.

I was surprised at his response, "Sengar, you must be naive to expect that I will believe your story. I can never believe that you did not do it on purpose."

I bid goodbye, saying I came to tell the truth, the absolute truth, as I leave the school, as I did six months ago. It was up to him to believe or not.

I saw three principals as a student in the school and have interacted with many later. All were good principals and worked hard but the man who transformed the school in every way was Wing Commander Sunandan Roy. Interestingly, he was not from the Education Corps. He was an ex-NDA fighter pilot, who had met with an accident and placed off flying. He imbibed the public-school culture and an officer orientation to the cadets. He established language labs, sent out students to compete in state and national events including national debating competitions, introduced proper dinner nights and many other activities. He managed plant machinery to convert orchards and hill sides into sports fields, a task we were doing manually for years. He revolutionised the way school teams trained and practised with good coaches. He introduced Western and Indian classical music and reading culture. He was instrumental in raising the confidence levels of the cadets and making it the envy of the best schools in Nainital and beyond. He was no educationist.

It is my strong belief that Sainik School principals must be a suitable old boy of the school, chosen by the Old Boys Association, in the same manner as the Rashtriya Indian Military College (RIMC), Dehradun. He could be put through a suitable orientation course on the management of an educational institute. The head master should be from the education corps to provide professional advice.

He introduced a voting system for selecting the school captain, but my love–hate relationship with the headmaster of the time, landed me one rung down. If he had his way, he would have me withdrawn, as I was on his mental withdrawal warning list even as a school vice-captain. My good standing with the principal was my firewall and I survived repeated assaults. Well, some of these were for silly reasons such as long hair.

My father visited the school to see how I was doing. He wanted to meet the head master. At a time when there was a cold war on between me and the head master, that was the last thing I wanted, as my father would have gone saddened and disturbed. That was the only time I lied to him to ensure that he did not meet him. I guess for a good cause.

Going to the NDA was the sole aim in life. I excelled in school at everything. I expected to do reasonably well in the NDA entrance examination. To my and everyone else's horror, I landed 272 in merit and the NDA used to take only 250 cadets. I was heartbroken and fell sick in the sorrow that I may miss the bus to NDA. I knew that the Army was what I was meant for. In fact, I would write my name on my school notebooks as Major Anil Sengar. A third generation, I was passionate and crazy for this profession. I was hungry to wear this Olive Green uniform.

God was kind; a month delayed, I got the happiest news of my life, a call letter to report to the NDA. Ask me, what happiness means. Ask me, what it means to be in seventh heaven. I thanked all those who chose not to join the NDA, it got me in, the last one in.

My merit of 272 hurt my pride. I knew that I was better than this. I gave no excuses. I just resolved to set the record straight in due course.

Will you believe that when I was in school, I knew that one day I would write a book on myself. For whatever it is worth, here I am.

With deep gratitude and admiration for the teachers and staff of Ghorakhal, who transformed our lives through their competence, commitment and devotion.

Forty years later in 2017, I met my classmate VP Singh in a wedding, who was the school captain in 1974–75. He said, "We all wanted to do what Sengar did in school, but we did not have the courage."

Lessons from School

While the whole academic environment has turned on its head, there are a few thoughts. It is disheartening to see the pressure on schoolchildren to get high percentages. Even children in lower classes are not spared the agony of competition. Sadly, some parents are accomplices in the crime.

My suggestion: childhood is the only stage when they will have fun, as they are meant to. Do not judge the children by your yardstick. See their perspective by becoming their age. Do not rest your prestige on their grades and on their behaviour. While you may want to live your dreams through your children, which is fair, it is a gross mistake to push them into a field they do not love.

The beauty today is that there are options for all kinds of talent to make a great career as long as you are good in that field. You do not have to be academically brilliant to make something out of life. Put your child in the field he wants to be, a job that he will love. You would give him a lifetime of happiness. There are no dearth of examples where students with poor academic talent went on to excel in other fields, simply because they loved what they were doing.

I started out as a weak student. I lived my childhood with gay abandon without a worry in life, had fun and laughed and played my part of boarding school pranks and mischief. Gradually, I excelled in everything. I think, I achieved all this, while having all the fun, for one simple reason—I had the realisation that I needed to make something out of my life, so never lost the focus. Arouse in your child the desire to excel, he won't enjoy the hard work unless he loves it and wants to do it himself.

Do not compare your child with others. God has made everyone different, and each brings his own unique bundle of talent: discover it, explore it, nurture it and let it bloom.

I had as good a childhood as one can get. It was fun all the way and I excelled as a student, all on my own. Make sure your children get one too. Do not hurry your children's childhood. It won't come again.

*

Thank You, Ghorakhal.

You gave us the diminishing values of simplicity and humility, honesty and integrity, respect and loyalty, hard work and fair play, empathy and compassion. All put together, you gave us what it takes to be a good human being.

* * * * *

The Days at the Academy: Forged in Fire

The National Defence Academy

The NDA is one of the finest institutions in the country. There are two things that make its unique character. The first is the perfectly laid out campus with awe-inspiring architecture of the eighteen identical squadrons and the imposing Sudan Block that literally towers over the whole academy. The awesome Khetrapal Drill Square with its surroundings is a sight to behold. The more important constituent of its character is the way of life and its traditions. Some of them sound silly, but they make the academy what it is. The NDA was a cultural shock to new arrivals. Those coming from Sainik Schools knew what was coming and weathered the storm easily. Nevertheless, the first two terms were a nerve-wracking experience that tested all your senses. The second termers who were the scum of the earth, just a few days ago, suddenly become the most dangerous sadists as if they owned the academy. Understandably, after six months of beastly existence, they assume the power to screw someone else's happiness and they do that with pleasure. The NDA has something that an ex-NDA never tires of visiting the academy, for nothing but to stand in the centre lobby of his squadron and relive the past.

For a first termer, the sixth termer is a lord. He never believes that he will get there one day. That looks too distant and improbable.

In the fourth term, I was marched up to the commandant, Admiral MP Awati. An impressive personality, with a not-to-be missed white beard, he was every inch like the ancient mariner. He spoke well with authority and excellent elocution.

As we stood outside the commandant's office, the buglers readied themselves to sound the bugle. That was the signal for relegation. My heart sank. Relegation for an offence you did not commit? I was not going to take that lying down. To my relief, the academy drill subedar major ran to the buglers before they could play the bugle and yelled at them, "Who asked you to play the bugle, *nikamme admi*?" I have never felt such a relief ever in life.

Three of us were marched up to the commandant. One of the three was Lt Gen Surinder Singh, the Western Army Commander in 2016–2017.

"I see that you are a Wing Corporal, and you are being marched up to me! Someone has to be out of his mind to appoint you one. You dare laugh when the deputy commandant of the academy is giving a talk. I will relegate you and put the term tallies of the previous term, do you understand?"

Relegation for no reason was something I dreaded. In fact, I would have quit the academy if I had got relegated. There was too much at stake to remain quiet.

"Request permission to say something, sir," I said.

"What do you want to say, corporal?"

"Sir, you have got the wrong persons here. It is true that some cadets were laughing when the deputy commandant was addressing the academy. But we were not the ones."

"Then do you know, who they were?"

"No sir, they were, I guess some fifth termers who were sitting a few rows behind, so I do not know who they were, but we were not the ones."

"I take your word, but whatever, I shall not accept cadets showing disrespect to my deputy commandant. Seven days' restrictions, march them out."

We heaved a sigh of relief as we were marched out.

A few months later, I was appointed the Flag Corporal. He is a kind of cadet ADC to the commandant. I stood at the auditorium to receive him for a function. As he stepped out of his car, I saluted him.

"Oh, you are the new Flag Corporal! Looks like I have seen you somewhere, have I?" He said in his resounding voice.

"Yes sir. I was marched up to you a few months ago."

"What! You get marched up to the commandant and some joker appoints you his corporal? Someone has a good sense of humour and you must be some character to be appointed one. Welcome aboard."

Awati was an impressive commandant who left his mark in the academy with new initiatives that gave the cadets wide inter-service exposure.

At the end of the fifth term, I officiated as the Academy Cadet Adjutant (ACA) but was dismissed from the deputy commandant's interview for selection of the ACA after a run-in with him on some issue. The squadron commander (Squady), Squadron Leader SE Ali Amir, called me to the office. "I know what has happened and knowing you, I am not surprised. I am glad you have the courage to speak the truth. I have kept the appointment of the Squadron Cadet Captain (SCC) for you."

He was a wonderful squady; firm, friendly and professional. In the fifth term I was a sergeant. I smacked a fourth termer with the sergeant's sash. His face swelled up and he reported to his squady, in Kilo Squadron. This squadron was in trouble for ragging at the time and he found in me the perfect scapegoat. He was bent upon getting me relegated and asked me to give it in writing that I had manhandled him.

I said, "I am not giving anything in writing unless I speak to my squadron commander."

My squady told me "Son, you have made a mistake. But it does not call for even a thought of relegation. That squady is trying to find some scapegoat for the troubles in his squadron. If you are relegated, I resign my commission. You are giving nothing in writing. Meet me in my office, I will be there in ten minutes."

There was no relegation. I was de-striped as a sergeant, to be appointed sergeant again in the third rotation.

Cadet Rider Sidhu, a General (twice relegated) in the academy was in my class. A happy-go-lucky character, he did not think much of the academy rules and regulations. He was in trouble with his squady and faced withdrawal.

He asked me, "Can you help?"

Surprised, I asked him, "How can I help? You better not get into any further trouble and you will pass out."

"Can you get me posted to Lima Squadron?"

I found that strange. I told him, "I am a cadet here like you, damn it. And I am not the son-in-law of the commandant. How the hell do I get you out to another squadron?"

Nevertheless, I told him I will try.

The next day, I walked up to the Squady Ali Amir. I told him the story and requested him to consider getting him to Lima Squadron to help him pass out. I took the responsibility for his good conduct.

Within a week, Cadet Sidhu was in Lima Squadron and passed out with his broad smile.

I salute Sqn Ldr Ali Amir. Such officers are rare. He conveyed a strong point: you give a man trust, you will receive trust in return. Sidhu lived by the book, because he was given trust.

Sidhu was a lively character. Nothing could dampen his spirits. He loved life no matter the circumstances. Sadly, he died young in a car accident. I wish to narrate one of his antics in the academy in his memory.

'Rider Sidhu', also called 'Pagal Sidhu', was a lovable guy and a good rider, hence the sobriquet. Sent to Echo Squadron on joining the academy, Sidhu soon realised he had to create his nuisance brand if he had to survive in the academy. There was one particular fourth termer bent upon making Sidhu's life miserable. Sidhu reali Sed he had to launch a counterattack for the long term good. Sidhu strained his brains more than ever in his life and had his plans ready. The next time the sadist fourth termer came to his cabin, Sidhu yelled, "You rascal, I will electrocute you and send you to hell!" and in a well-rehearsed move, yanked out the table lamp with a broken bulb and lunged at the fourth termer. The fourth termer ran for his dear life with Sidhu on the chase, lamp in hand. The cat-and-mouse chase continued as the senior, yelling for help, ran from one floor to another. Sidhu, allowing him to maintain the gap, kept shouting, "Bastard, I will electrocute you!" Hearing desperate cries, his coursemates ran out of the cabin and came to his help. Sidhu was taken under control and paid for it. Thereafter, he earned himself the sobriquet of 'Pagal Sidhu' (mad Sidhu). Sidhu's reputation spread like wild fire and that was the end of all his woes in the NDA forever. No senior ever again took the chance of getting electrocuted by a crazy surd. With all his problems over, Sidhu was happiness personified. With no worries in life, he enjoyed every single moment in the academy. Already a 'General' (relegated twice) by the fourth term, he had a tremendous attitude and zest for life. Together, both he and his brother could have easily qualified for the Guinness Book of World Records for becoming 'Generals' in the shortest time.

Sidhu had his fun with juniors but never harassed them, as some others did. Amongst his many original accomplishments was a 'musical orchestra' that he had created. It consisted of eight first termers whom he had trained as musical instruments. Each of them was trained to produce different types of musical sounds varying from tabla, keyboard, drums, guitars, etc. He had trained them to play in two modes, the auto mode and the manual mode. To activate the orchestra, he would poke his finger in the stomach of the first in the line, who would then twist the ear of the one next to him and so on. The last one would produce a ping to indicate that the orchestra was all tuned and ready. The orchestra would be lined up by Sidhu in a musical logic that only he understood. Every Saturday evening, he would organise a musical evening by his band on the top floor flank of Lima Squadron. Mostly, the captive audience consisted of first and second termers, even though he had a good course fan club too. He would produce different tunes by poking his finger in the stomach of the human instruments in a well-rehearsed pattern, and incredibly, produced good music. His captive audience, as instructed by Sidhu, clapped in deep appreciation of the performance after every tune. On the auto mode, activated by twisting the ear of the drummer, the orchestra would play 'Colonel Bogey' and 'Auld Lang Syne', the popular tunes on the drill square.

Squady Ali Amir was just the right kind of officer and Lima Squadron had an excellent environment of growth under him.

As I said, I was way down in the entrance merit to the academy, fortunate to scrape through into the NDA with a very low merit of 272. Those waiting days were the worst days in my life. All that I wanted for a career was to be a soldier. From early days in school, I would write my name on the notebooks as Maj Anil Sengar.

The merit of 272 was a judgment on my capabilities that I was not prepared to accept. I gave myself no excuse and blamed no one. Clearly, I had not performed to my potential. In the second term, when there was some time to ponder, I decided to find my rightful place and set a goal for myself for the Indian Military Academy and achieve that honestly and fairly.

I was a fitness freak in the academy and was known as Tarzan of the course and later christened the 'The undisputed King of ropes' during the course get-together by Commodore Sanjay Thatte. On Sundays when the academy slept, I wore my packs with bricks and went for long runs and climbed ropes. I raised my mental robustness to a level that I knew no fatigue. I learnt a lesson that the human body's capacity to deliver is only limited by what you can bring your mind to demand. The body will rise to any level. This was to come in handy at the Indian Military Academy.

My coursemates remember me for rope climbing and legendary fitness. When I released my second book, someone commented, "I thought Sengar was all about rope-climbing and muscles."

It was my daily ritual to go to the academy gym and climb the rope first and second class over thirty odd times, and eight hundred push-ups and sit-ups each.

As the SCC, my study periods were used for workouts. One day, as I pushed myself, push-up after push-up, Flt Lt Razzaqui (later Air Vice Marshal Razzaqui), a Divisional Officer (Div O), opened my door. I looked up, drenched in sweat. He stood there for a while, quietly closed the door and walked away without a word. He was an intellectual kind. Transistors were banned in the academy. He once heard a transistor play, but walked away because the cadet was listening to news.

One day, a PT instructor walked up to me.

He said, "So you think you are very fit? You want to have a competition with me?"

I said, "No, I do not think so. I just love keeping fit and rope-climbing, so I come here every day. You are our guru, and I can never compete with you."

For some reason, he was bent upon taking me on. I said, "If you wish so, we can try it out."

He asked me "What do you want to do?"

I said "Anything; you choose—beam, push-ups, sit-ups, rope, anything in this gym."

He chose vertical rope. The competition – climbing the rope first class fastest, touch the ring and come down.

By the time, he had come down, I had gone up and down four times and was half way up the fifth time without touching the floor.

He looked surprised and said, "Cadet, you are good" and walked off.

What the academy taught us was camaraderie, espirit–de-corps, pride, devotion to your organisation, mental robustness and an uncomplaining attitude. In all our six terms, we had one lone quotation on the study table, 'Cribbing is Unofficer-like'.

It is quite amazing that an ex-NDA never tires of visiting his squadron. When one ex-NDA meets another ex-NDA, the first question is which squadron, which course, and the deal is done.

The nature of warfare has turned on its head over the last four decades. So has the socio-economic fabric of the country with its impact on the intake, both at the officer level and the men. The academy must constantly evolve itself to meet the challenges of the future. The leadership needs should be honed to adapt to the future needs of the conflicts and the changed nature of human beings that are a result of the changing socio-economic scenario.

Colonel Dalvi, a veteran is spearheading the 'Victory India Campaign', addressing important issues that trouble the military by collecting views from experts, both military and non-military. He has carried out a thorough analysis of many institutions including the NDA with pragmatic suggestions for the future. I hope that the concerned hierarchy finds the inclination to read and study his report and implement it as appropriate.

The NDA has a great tradition called the end of term social. There is nothing social about it. On the contrary it turns out to be most anti-social and sometimes brutal. The sixth termers call the shots for five terms. Some of them are real nasty with juniors and make their lives hell. This social is an opportunity for the juniors to settle scores with the sadists. The sixth termers on this day place themselves at the disposal of the juniors. They serve the juniors tea and run their errands. The sadist variety would get thrashed pretty badly, but they bore it with courage.

The academy witnessed an ugly incident in the Lima Squadron during the 48th Course passing out in Jun 1975. As was the academy tradition, the sixth termers were banged up by the juniors during the end of term social. The Cadet Sergeant Major, who was apparently quite a sadist, was banged up brutally and he died of his injuries. Another cadet was severely injured and had to be medically withdrawn.

The Lima Squadron was disbanded and reconstituted from the rest of the academy. The proverbial 'muck', who landed there had no outdoor skills, the only thing that matters to the squadron. I joined the squadron immediately after that in Jun 1975. For the next five terms, Lima Squadron did not win even one medal, individual or team event. That was quite a shame. I eventually became the SCC and we got a new Squady, Sqn Ldr Jadhav. We decided that this state of affairs was not acceptable and we got to earn our spurs. The first inter-squadron competition was the cross-country. The squadron had heard such talks before, but not much had been achieved.

The NDA cross country is an event in itself. About two thousand cadets in multicoloured squadron vests line up in the unending glider dome. Last minute motivation talks by the SCC, yay-ya Lions, and then the whole academy taking off as if there is no tomorrow, is a sight to behold. Add to that scores of officers smartly mounted on well-bred horses cantering along. How strongly the cadets wish that they were on horseback too, but there is no getting away. The cross-country route was pretty taxing with the 'Rape Hill' adding salt to the wound. The squadrons worked very hard to do well, this invariably being the first competition of the term.

As the SCC, I had worked bloody hard-running during games with the squadron and again after dinner with weak runners. I was confident the squadron will do very well. On the day of the competition the squadron fell in before moving to the glider dome. I pepped up the squadron, saying that the time had come to put in their best and deliver. I said that everyone had to cross at least one cadet in the last hundred metres. One second termer tried to put me wise saying, "How does that matter sir, it's enclosure system so you get the same points whether you come first or last in the enclosure." His logic was irrefutable. I told them this story why every bit of extra contribution was important.

"There was a surd, a jat and a four-feet-nothing chink from the hills who went trekking in Africa. They lost their way and strayed into a cannibal tribe territory and soon were caught. They were brought to the tribal Chief who was up in rage. The Chief ordered a bonfire with the whole village in festive mood performing tribal dances. He ordered a feast with one man being barbecued at a time. The three musketeers were trembling with fear, certain their time had come. They fell on the Chief's feet and begged for mercy, pleading that they had made a mistake and would never be seen around ever in Africa. The Chief decided to give them a chance under one condition. He said that their 'dicks' would be measured the next day at dawn. If they measured twenty inches or more then he would spare them or else they would be eaten alive. Having no option, they agreed. The night passed, somehow. Next morning the whole village had collected at dawn. Three of them were stripped naked and taken in front of the Chief. The Chief ordered the measurements to be taken. The surd measured nine inches. The Jat measured ten. Pretty close, they were certain they would be through. The chink from the hills stepped forward and measured one inch. Twenty inches, declared the chief's man. Three of them exchanged glances with great relief. The Chief, as promised, released them under one diktat. They had one hour to disappear from his territory. He would send his men after them and if they were caught, there would be no mercy. They ran like hell for two hours, and when they were certain that they were out of his territory, sat down to catch their breath. The Jat claimed all the credit for saving their lives. The surd said his contribution was no less and claimed equal credit. The chink kept listening silently. He then said, "You are claiming all the credit for saving our lives. Mine is the most important contribution in saving your lives." Both the companions laughed and asked him how his contribution was the most important? The chink said in chaste Hindi, "*Agar aine mauke par mera khara nahi hota na, toh tum shaare mare jaate.*" (If I had not had erection at the critical time, you would be all dead). "So Lions as you can see, every inch of contribution is crucial," I said.

The squadron in high spirits responded by a loud yah-yah. The Lions were convinced by the story, and when the cross-country competition was over, the squadron had finished a disappointing eleventh out of twelve squadrons! So much for the moral lecture. I was heartbroken. As for the result, can't really blame the squadron. It was probably the only time in the history of the NDA that there was a re-run in the academy. In that maddening commotion of the last hundred metres' rush, the third enclosure types, wanting to get their squadron extra points, had gate-crashed into the second enclosure leading to a re-run of the whole academy less the first enclosure types.

I was terribly disappointed. The squadron had worked real hard and I had given it my all—my heart and soul. My introspection led me to the conclusion that I had pushed the squadron too hard and they were all fatigued and tired and many suffering from stress fractures by the time the competition took place.

A day after the competition, I was called by the squady. I thought it was some routine issue, as the SCC meets the squady on regular basis.

As I stood in front of the squady's office, I saw all the Div Os sitting there and the squady sitting at his table in a pensive mood. It was eerily silent. I realised, something was amiss.

Come in, the squady said, very seriously.

I marched into the office.

"Sengar, it is disappointing that early in the term, we have to call you here on an unpleasant note. You know the result of the cross-country competition. The squadron has come eleventh out of twelve squadrons. You have been living in the squadron like a king and sleeping and doing nothing. You have failed to discharge your duty and we have decided to de-tab you and relieve you of this appointment."

Well, that was a rude shock. I had given the squadron everything for this competition. And to be told that I had been sleeping was too much to digest. A grenade exploded inside me. I replied instantly—that for sure was very rude as well.

"Sir, you are right. The squadron has performed badly. I am the SCC and I take full responsibility for the poor performance. I shall give no excuses. But, I take offence to your judgment that I have been sleeping and I have done nothing. I am sorry, if you say that I have been sleeping, it is clear that all of you sitting here do not know how hard the squadron had worked for this competition. The academy is surprised that the Lima Squadron has fared poorly because they know that we gave it our all. Sir, do you know that I ran during the games time with squadron and I ran with the weak runners after dinner. The fact that I am the only one from the squadron in the first enclosure, indicates that I could not have been sleeping. As I now realise, we fared poorly because I pushed them too hard and they were already tired and fatigued. It is not the lack of effort. On the contrary, poor performance is a result of trying too hard. Sir, none of you ever came to see the squadron practice and if none of you know what the squadron was doing, clearly you all were sleeping and not me."

That outburst shocked them. None expected a cadet to talk back to the squady so ferociously.

"So, you are trying to tell me that we were sleeping?"

"Yes sir, if you do not know how hard we worked and tell me that I have been sleeping, surely you are not aware of what was going on in the squadron."

"Get out of the office. You will get my directions," he said.

I felt hurt at the allegation, when I had given my heart out. The next day, my Div O, Captain Tully, Navy, called for me.

"You clot, is that the way to speak to the squady? Are you out of your mind?"

"Sir, I am sorry for the manner I spoke. As I said, I gave the squadron my heart and soul for this competition. Each cadet gave his best, but as I said, I pushed them to my standards and they were fatigued and tired. To be told that I did not do my duty, touched the wrong chord. It hurt my pride. As for the sense of what I said, I stand by what I said."

"Squady is deeply annoyed and he has decided to de-tab you as the SCC."

"No issue sir. I had never asked to be made the SCC. If he thinks someone else can do the job better, I will be happy for the squadron."

I was not de-tabbed and the squadron performed impressively on all counts.

Lesson Learnt

The results are not always commensurate with the effort that you put in. Sometimes, it is because your approach is faulty as was mine. A well-planned action plan is necessary to achieve proportionate results. While what the Geeta says, that your duty is to work without aspiring for rewards, is true and a secret of happiness; in this environment, it is not a bad idea to apprise your superior commander of what you are doing.

The next lesson is that as a leader you may have very high standards. While that serves as an inspiration, pushing everyone to your standards may be counter-productive. It is important to keep in mind individual limitations and empower them to surpass their best.

Three decades later, when I was the divisional commander, my habit of working without selling myself stayed with me. I did not speak to the corps commander for over four weeks after taking over. He called me up and mentioned my long silence and asked what is happening in the division. I said, training, training and training. I said when you visit my formation you will know what we are doing. He said, I must call him sometimes to keep him apprised of what we are doing. That says it all. It keeps the superior at peace.

The Indian Military Academy: Swimming Against the Tide

The Indian Military Academy (IMA) was a come-down from the NDA in every way. It was great fun nevertheless with all kinds of entries in the fray, each with their own character.

IMA repeatedly suffers from becoming a parking lot for the sons and sons-in-laws of senior officers, preparatory to staff college examination. IMA is too important a place to allow such things. The quality of officers that get commissioned into the Army will be a direct reflection of the instructors there, both officers, JCOs and NCOs. It is a long-term investment in which there can be no compromise. I have heard a commandant complain of the deputy commandant not being committed enough and not doing enough. Such important postings cannot be parking place for anyone. Such postings must be based on spoken reputation and not on paper record of service.

As I said, I was not prepared to accept my NDA entrance merit as a judgement on my ability. I knew I was better than that. I had set my goals for the IMA. I worked hard in the IMA—honestly and fairly. I received the sword of honour and the gold medal, the aim I had set for myself in the second term in the NDA. That was an achievement that I have always been proud of, for it was against great odds, and the fact that all the medal winners of my course in the NDA were outstanding army cadets. I was swimming against the tide even before I had reached the academy.

It is therefore necessary to narrate my experience in this context, with the hope that this is not allowed to be repeated.

I was competing against unfair competition on two counts. One was a cadet whose father was a senior army officer posted in Dehradun. He tried all affiliations and tricks to help his son to get to the top. Nothing unusual in this, I guess. Interestingly, he was more obsessed with monitoring my performance, hoping I was screwing it up somewhere in exercises.

It was the second that was disturbing and a bigger challenge.

During our time in the academy, there was an instructor from a particular community. An officer from the ranks, he manipulated to get good cadets of his community coming from the NDA posted to his platoon/company and help them unabashedly against all norms. He did that in my course too. He would manage and give him all the test papers a day before the exams. Given all the marks and influence that a platoon and company commander carry, I was competing blindfolded with my hands tied behind my back. An unfair competition unworthy of a military academy.

How do I know this?

This cadet either did not have the intelligence or the will to put in the required effort to answer the question paper. So, he chose a 'beast of burden', an intelligent friend who would solve the paper for him, himself becoming a big beneficiary in the process. And there was a third cadet in this group, also from the chosen community.

This intelligent friend and I landed in the military hospital, Dehradun, due to injuries and spent a few days on the neighbouring beds.

He told me this story.

Why did he part with this secret? Because his friend having gotten into a spot due to immoral acts, surreptitiously tried to spread the disease to them too. He had stooped too low to save himself that could have got them withdrawn too, revealing his true character. These guys were in a rage.

This instructor would quietly take this cadet out on his scooter for private treatment at night. The story is too sordid to be reproduced here.

How is it that a cadet who is relegated in the NDA on moral turpitude gets graded with best 'Officer Like Qualities' in the IMA? Maybe they turn over a new leaf, maybe there is something more.

Parochialism is a malaise that will destroy this wonderful Army. I strongly believe this malaise is afflicting the Army. It happens on caste basis and regimental basis. I do hope this stops in the interest of meritocracy.

As if this was not enough, God, if there is one, tested me on another plane. The IMA has a competition for best physical fitness for the award of 'Para medal'. I was a strong contender. One day while practicing the Russian Obstacle Course, for the Para medal, I landed on a stone and broke my ankle. As fate would have it, it was just four days before the all-important event called 'Camp Chindits', a gruelling week-long camp that tests your nerves as much as your body.

I was faced with a great dilemma. I was the Academy Under Officer, and I knew I was leading the course. If I reported sick, I would land up with a plaster and miss the camp and lose marks for this major activity. With that would have evaporated my dreams of the sword, a goal I had cherished and worked for three years. There were enough hawks monitoring and waiting to pounce on me. I had invested everything for this one goal. I was not giving it up, having come so close.

I did not report sick and decided that I will do the camp, no matter the pain or the cost. Believe me, I did not take off my left boot for eleven days, four before the camp and seven during the camp, out of fear that if I took off the boot, I will not be able to pull it back on the heavily swollen foot. I managed the four days before the camp hobbling on one leg and weathered the torture of Camp Chindits, as the worst week of my life, hobbling on one leg with the support of a stick through the nala beds and the rocky hills. When the instructor learnt of it, and he did it in no time, he said, "You must be crazy to think that you can go through this camp with a broken ankle." I was reminded of this by two of my coursemates four decades later.

I am proud, I did. I am equally proud with all humility that I did not allow these cadets to come close despite an unfair game.

The problem was not over after the camp. As the other contenders realised that they had lost the race for the sword, they lost all interest in commanding the Passing Out Parade rehearsals. It then fell upon me to command the parade every single day for over three weeks. The trouble; with a broken left ankle, I could not stamp my foot. As a parade commander, all your movements are in isolation and the drill is all about smart movements and impressive stamping sounds. It was a testing time and I conquered the physical pain by the strength of my mind and willpower, something that I had built in the NDA, pushing myself to the limits and beyond.

The IMA is a great place. A conglomeration of a variety of entries, it has its share of humour, every step of the journey.

Gentlemen Cadets take immense pride in putting up a good show in the passing out parade and practice hard for it. There are days when they just want a break. Like any other thing, there are days when nothing goes right, however hard you might try. During one such day when Major John Taylor, the Adjutant, was present for the rehearsal, nothing seemed to be going right. The turns were atrocious, swinging and digging terrible and the rifle drill disappointing.

Despite his best efforts to egg the cadets on, at which he was good, nothing improved. He said, "You do one good salami shastra (salute with arms) and I will break you off." Thinking probably that a change in the word of command will make the difference, he told the drill Subedar Major to give the word of command. Typical of the drill Subedar Major, he egged on the cadets and after a brief pep talk gave a full-throated word of command 'Salami Shastra'. The cadets went through the drill motion. As luck would have it, it was worse than before. There was a lousy rumbling sound of thumping of rifle butts and stamping of feet as opposed to the crisp crack of front guard and slapping of the rifle butts. Everyone knew it was a horrible drill movement.

As the lousy drill movement commenced, John T., slumped on his horse, head bowed down in disappointment. And, just as the drill movement finished, the Adjutant's horse shit a wholesome, accompanied by a mighty fart that was heard across the drill square. There was pin-drop silence as the cadets waited for John T.'s reaction.

John T. was not the one to miss such an opportunity. He rose and stood tall on the stirrups, turned fully backwards from his waist, looked at the rear of the horse, turned around, surveyed the parade from left to right and yelled with bewilderment, "Look-look-look, like me, that is what even my horse thinks of your drill. *It is all bloody shit.*" He said "break" off and rode off on his elegant steed.

John Taylor was an excellent Adjutant; he did his job to the hilt and kept the cadets amused, getting the best out of them. Havaldar Swamy was a drill instructor par excellence, straight from the Marine Corps films. He worked hard, squeezed the last ounce from the cadets and earned respect through his honest devotion to the academy. When the cadets passed out, he would position himself suitably and salute and convey his best wishes to every commissioned young officer, whom he was ready to skin as their instructor just a day before. Great specimen of a soldier.

The IMA has to train cadets to lead men to succeed in battle at the lowest cost to life and property. It is a multifunctional responsibility. Yet, the most important responsibility of the academy is to train the cadets to be men of character and integrity, the foundation of good leadership. Good instructors only will produce good young officers.

When a cadet walks over the 'The Final Step', the thirty-inch step is giant leap that transforms him from a simple Gentleman Cadet to an officer with immense stature and responsibilities. Three things happen to him:

- As an officer, he assumes an unlimited liability to serve the nation, the army and his men. That means sacrificing your personal interest in a manner unimaginable in the civil world.
- The second, he earns the privilege to wear the uniform and that brings upon him the unquestioned trust and the faith of the people of the country. And that means that he is under obligation to live by the highest standards of professional and personal conduct on and off parade, in and outside the cantonments. All soldiers must remember that they are the ambassadors of the Army. When they step out of the cantonments, for the people of the country, they all wear one name plate and it says INDIAN ARMY. Every act of commission and omission affects the image of our Army and we owe it to the martyrs of the Army, that we play our part in upholding the dignity of the uniform and image of the Army.
- The third, he becomes a combat leader. A great transformation from a humble Gentleman Cadet, whose only responsibility is to pass out of the academy in one piece.

Every young officer is just ripe to be nurtured into an excellent officer. What he needs is a good start and a good grooming early in his career. That should be the prime responsibility of a commanding officer.

Lessons

My advice to the young generations: It is your right to dream and have ambition. It is your right to work bloody hard to achieve that ambition and compete. But there is huge satisfaction in doing that fairly and honestly. The recognition that will come for doing so will be a source of immense happiness for long. It will earn you the reputation that is invaluable and give you a reason to walk tall. Remember, if you can be genuinely proud alone in a room, you are okay. Success based on cheating and deceit has a short shelf-life and often embarrassing. Both these characters in my course never lasted long in the army, even though they finished in the top bracket of the course.

The lesson here is, if you have a dream that you are prepared to die for, a dream for which you are prepared to sacrifice your time and life for, a dream without which life does not seem worth living for, a dream for which you are prepared to say that I am not done till I have won, believe me, it will happen. It has happened for me all through my career but it has never come easy. It took sacrifices, time, effort and commitment.

Those who say they can and
those who say they can't are both generally right.

You cannot choose your circumstances, but a man has the power to rise above his circumstances and thus change them. It will happen only if you are hungry to achieve what you aspire to achieve.

Nothing comes easy. It becomes easy when you are crazy about it. It becomes easy when the dream that you dream prevents you from sleeping. I was crazy about my dreams and had already received the sword in my imagination a hundred times before I actually got it. Well, I did suffer a permanent damage to my ankle, but that was a price I was prepared to pay and was worth it.

The best compliment on my achievement in the IMA came from a coursemate Colonel SK Attri, an aviator of great competence, forty years later. He sent a message on the course WhatsApp:

> *"I remember how you did not remove your boots for days after breaking your ankle on the obstacle course and did the Camp Chindits with a broken leg. You were my belief that good also survives in this world."*

That makes all my trials, tribulations and pain worth it. I was surprised that he remembered it four decades later. I had forgotten it.

EXPERIENCES ACROSS THE SPECTRUM

The Army is a service where team work is the key to success. In the extremely intense cutthroat competition of the Army, I have seen the best, not even sharing notes on the course, even with friends. There are also the other types, who will be happy to ride on the thrust of others but will not share anything with others.

As for me, I have never hidden anything from anyone. Anything that I had was available to anyone who thought it was useful. I knew, no one can excel on the power of notes alone, it needs a lot more.

At the end of the commando course in early 1980, Lt Rajiv Sharma, (Mech-1 JAT) confessed to me that four of them had ganged up to beat me in the course, taking turns to tire me like the pace setters. They failed. I appreciate his candour in confessing.

Well, the commando course had great moments of tragicomedy too. The toughest exercise in the commando course is the forty kilometres speed march, the last event of the course. By then the body is breaking down and the mental and physical states are at the breaking point. This is conducted after the survival exercise, where you live off the land for forty-eight hours. In the middle of forests, nowhere near a civilisation, there is little food that you can smuggle through with all the ingenuity at your command.

After forty-eight hours of living off the land, trudging through rivers and nalas, you are very-very hungry and almost at the end of your mental and physical endurance. You are given a hot meal at three in the afternoon and the speed march commences shortly thereafter.

Two days without a meal and with the monster of forty kilometres speed march looming, there is little appetite for food, and I guess, one is too hungry to eat. A commando carries a twenty kilo sand bag on his back which is weighed both before and after the speed march. The distance of forty kilometres is required to be covered in six hours. Considering that the commandos have had little to eat and almost no sleep in the past two days, and with every passing day sapping your strength and endurance, it is a tough undertaking by any standards. In the old days, it was mandatory to pass this test for an 'instructor grading' only. While everybody started the speed march, many, not in the run for an instructor grading or up to testing their endurance, would take advantage of the rule and sham it out. Since this exercise has a high rate of dropouts due to heat and dehydration, there is a 'STRAGGLERS' VEHICLE', a 3 Ton, with a big board in front and rear that trails the commandos.

Thirty-five days of physical and mental battering takes its toll. A commando by this stage is surviving purely on willpower. Those who fall off or cannot continue the speed march are picked by the straggler vehicle, which also carries the necessary medical cover. With us on the course was quite a character, Sangha. He considered the speed march a human rights violation and had no appetite for such disruptive things of life. Waiting at the start point before the speed march, I overheard him telling his friend Dallal from the Punjab Regiment, "*Guru, speed march shuru hote hee straggler vehicle pakar lena, baad mein pata nahi jagah mile na mile.*" (Partner, get on the straggler vehicle the moment the speed march starts. You never know, you may not find a place later.).

And believe me, it was less than half a kilometre from the start point that I happened to look at the straggler vehicle following us. I saw Sangha and Dallal in the business class seat of the 'Straggler Vehicle'—the co-driver's seat in front. They took no chances and hopped in even before the vehicle had crossed the start line.

I love everything about the Army, however stupid. There is never a dull moment. This same Captain Dallal cited above narrated this incident to me, a real incident of blind obedience. It demonstrates in its own way, tragicomic though, what makes the Army an organisation that delivers by virtue of military discipline and blind obedience, even if silly sometimes.

The Army works on two rules:

Rule 1—The senior is never wrong.

Rule 2—If the senior is wrong, see Rule 1.

The 'senior can do no wrong syndrome' has taken deep roots in the Indian Army. Stupid, it might appear at times, but the Kargil War could not have been won but for strict discipline and blind obedience.

Captain Dallal, 'a happy go lucky' bloke was the adjutant of a Punjab battalion on the 'Line of Control' in Jammu & Kashmir. Gifted with bloodshot red eyes and a muscular build, he looked drunk twenty four seven-any time villain looks. He was called by his commanding officer and told that Brigadier Raja Ram from the Corps Headquarters would be coming to the battalion for area familiarisation and would be having lunch in their mess. No problems, said Dallal, the Brigadier was his instructor in the Infantry School during the Young Officers' course and he knew him. His commanding officer added, though he did not understand why, that he was told to keep an inflated scooter tube handy for the Brigadier in the battalion.

Dallal like a good adjutant managed two inflated scooter tubes from a scooter repair shop close by-one plus one reserve. Dallal and his boss waited for Brigadier Raja Ram discussing what the tube was all about, failing to find an answer.

"Surely, he couldn't be having plans to float across Manawar Tawi into Pakistan," thought Dallal.

Two military vehicles streamed in raising all the dust in their wake. Out came Brigadier Raja Ram with an inflated scooter tube in hand. He met the commanding officer and met Dallal, warmly recognising him as his old student. Dallal, a simple Texan had an inflated tube in his hand and kept wondering about the name of the game. Fifty yards behind, hiding behind a bush was a burly Sikh jawan, waiting with the reserve tube ready to appear on a pre-decided field signal.

Brigadier Raja Ram then handed over the tube to his driver. Dallal followed suit and handed his over to his runner. The team was then conducted around the battalion defences and brought to the mess for lunch. As Raja Ram approached the mess, his driver came running and handed over the tube back to him. Dallal followed suit and took his tube from his runner, wondering what he would do with it inside the mess. Brigadier Raja Ram entered the mess, placed the tube on the sofa and sat down on it. Here on, the episode is reproduced verbatim as narrated by Dallal himself.

"Seeing him sit on the tube on the sofa, I thought it was some new custom of sitting in the mess, which we had not come to know being in the field. After all he was a Brigadier; he knows what he is doing. So, I also placed my tube on the sofa and sat down on it.

Seeing me do that, Brigadier Raja Ram said, 'Oh! I didn't know you also had piles, Dallal.'

As he said that, my commanding officer and other officers looked at me, rather embarrassed at my doing.

Then I realised why he was going around with a tube. I sat there quietly; embarrassed, sheepish and shattered not knowing what to do with the tube under my butts. All the officers were served beer. When my beer came, much less drink, I wanted to drown myself in it for my stupidity in following him blindly. It took me many days to overcome the embarrassment."

So, life can be funny, stupid and tragic all at the same time.

I have talked about my father's philanthropic nature. We seemed to have inherited these traits too.

There are two things that multiply by division; knowledge and happiness. In the Army, when you help someone, you only invest your time and nothing else. The rest belongs to the Army, yet we have people who find ten reasons to say no, and not one to say yes.

I was doing the pre-staff course at Pune, preparatory to the Staff College entrance exam. Major Gurinder was with me in the same syndicate. During a tea break, early in the course, he came up to me.

He said, "I am Gurinder from 5 Grenadiers. You do not know me, but I know you as you were the Academy Under Officer in the IMA when I was a first termer. I have come up with a request. Please feel free to decline, if you wish. I believe that ninety-nine percent officers in your place would decline, so it is fine if you say no."

I said, "What can I do for you?"

He replied, "No one from my battalion has passed the staff college exam in ages. I also have no chance in hell passing on my own. I have observed you in the syndicate for the last two weeks. I feel if I study with you for two months under your guidance, there is a chance that I may pass."

I had never seen that officer before the pre-staff. Before he had finished, I told him, "I will wait for you. There will be accommodation arranged for you next to mine in the unit. If you think I can be of help, it will be my pleasure."

A complete stranger, he stayed with me for two months and cleared the exam.

When I was doing the staff college course, then Colonel Mohan Pande, Mahar Regiment, was our syndicate Directing Staff. Suddenly, one day, out of the blue, he asked me,

"Are you the Sword of Honour of your course?" I said yes.

He then said, "Were you in the IMA basketball team?"

I said yes.

"So you joined the basketball team to get some extra points for higher merit?"

Ten years after I had passed out from the academy, and clearly of no relevance, I was surprised to hear this from nowhere. I asked him,

"Who has been telling you all this?"

"No one, I just asked."

I said, "Sir, you won't know it unless someone told you. The slant of your statement is clearly derogatory. Have you asked others who were in some academy sports team? Is there something wrong with that, then why have the academy sports teams?"

"No, I just mentioned," replied Colonel Pande.

"No sir, it is not that simple. It is someone here in the course who is trying to discredit me and your entertaining frivolous talks is indeed surprising."

The moral is, watch out, your success will please very few, if it does at all. Many would continue to waste their energies and time in pulling you down rather than in pulling themselves up. Lots of people will discredit you if they can't defeat you.

In the context of this episode, far too many people are losing their sleep because someone else is doing well.

My brother, Col Ashok Sengar, a paratrooper par excellence, was known for his great heart and mind.

I was posted in Infantry School, Mhow, in 1990. He had driven to Mhow, and on his own, offered me his car for the staff college course at Wellington, because my two kids were less than eighteen months old. A bachelor owning a small car, he had plans to drive along the coast spending a month on the way.

Sitting in the college bar, Colonel Subramanian from 4 MADRAS, also posted in Infantry School with me, got chatting with him. He learnt he was there to drive the car thousand miles to Wellington for me. Subramanian then made a strange request. He said that his father was sick and he needed to be at Bangalore at the earliest, and at this short notice, he was not getting a reservation anywhere.

"Can you give me a lift?" asked Subramanian.

"Lift to where?'

"To Bangalore?" replied Subramanian.

"How many persons?" he asked without blinking an eyelid.

"Four in the family", replied Subramanian.

He drove this family straight to Bangalore, over a thousand miles, cancelling all his holiday plans. A man he had never met before and a man he would never meet again, he had nothing to take from Subramanian.

How many people do you know who will do this for an unknown man, and for nothing; will you?

These are great qualities of a leader and a human being. Is not the Army lacking, where such qualities of character and heart count for nothing in the promotion system? No wonder, he was known as raja saab, amongst his soldiers. A man with a big heart.

Three decades after the commando course, I went to the MS Branch and met a coursemate, exploring the possibility of a posting to New Delhi, as I intended to quit, having missed the board to Brigadier. My coursemate who later became an army commander, said,

"I must help you, for one simple reason. In the thirty kilometres speed march in the commando course, I had suffered a heat stroke and was throwing up. You were leading in the course and heading for the commando dagger. Despite my telling you to go, you refused to leave me, risking failing the speed march."

Nice of him to remember, I did not even remember it remotely.

Well, he took me to Colonel MS 1, who later became an army commander. Having heard my request, he saw my record of service on the computer and said that I cannot be posted to New Delhi. I asked, why not?

"You are too high profile to be posted to Delhi."

Wow, flattering to hear, I found that amusing.

I said," The army finds me unfit to be promoted to Brigadier. And you are telling me, I am too high profile to be posted to New Delhi. Even otherwise, do high profile officers not get posted to New Delhi?"

Cornered, he continued, "No, I do not mean that. You are a superseded officer and no one wants a superseded officer in New Delhi."

I said, "Fine, are you trying to say that there are no superseded officers in New Delhi?"

"Do you have any other excuse?" I asked him and left the office.

An officer with this attitude—an army commander! What do you expect from him?

I always went out of the way to defend the deserving officers, irrespective of the opposition.

I was posted in Turkey. One of my excellent officers in the unit gave me a call to tell me that he has been overlooked in the promotion board to Lt Colonel. I was genuinely shocked, as he was an excellent combat leader with a great attitude. He told me that the divisional commander has reported that the commanding officer had shielded him by keeping him in staff appointment during the 'Operation Parakram' mobilisation. I asked, "How do you know this?" He said, "Do not ask how. I have the photocopy of my report written by him."

I asked, "What can I do to restore justice?"

I gave him a two year backdated letter of appreciation on behalf of the divisional commander who had fixed him. The letter did the trick. The divisional commander, based on my letter of appreciation on his behalf, possibly realised that he had unjustly reported on him as a case of mistaken identity. He got his redressal.

For an officer to have fun, it is important to develop the attitude to love the job that comes your way, to do it to the best of your ability and to hell with the rest.

I was posted in the Weapons Wing, Infantry School, Mhow, in 1989 as a Major. The Weapons Wing stages an infantry fire power demonstration for all officers attending courses in Mhow. This is a prestigious event and conducted under immense pressure. The commander called me and said, Lt Col X who was detailed as the conducting officer had to leave due to a personal emergency. Can you take on this important responsibility at short notice?

I said, why not? If someone can do it, so can I.

We practiced just one day. Lt Col Pathania, AMC, (later Maj General), who did the staff college with a competitive vacancy, was providing the medical cover. After the main show was over, he said:

"Partner, I have seen many of these fire power demonstrations. I have seen people shitting bricks here. I did not see one frown on your face. After one briefing, you never said a word, and everything went perfectly. How do you manage to retain this calm and poise?"

I smiled and said, "People get worried over inconsequential things. The worst that would have happened is that some firer would have missed a target and that earthen pot would not have shattered. And that is not earth shattering. The best firers, even in the Olympics, miss their mark. These are no Olympians. No sweat at all. And if someone feels that missing a hit is an unpardonable crime, he can deal with me."

He said, "I have never seen such a cool attitude. You will go a long way. See you as an army commander one day."

I missed my board from Colonel to Brigadier!

Posted as the officer-in-charge {GSO(I)} Training in the GUARDS centre, I was surprised to have the Deputy Commandant of the Centre come to my home in the office time when I was on leave. There was a long pending project of writing three training precis for recruits under the training battalion commander. This project had been in a mess for years, getting nowhere. The previous day, the responsibility for this project was given to the deputy commandant on posting out of the training battalion commander. That was a rude shock to him.

He said, "You know this case. The commandant has given it to me. There is no way I can do this. If there is anyone who can do it in the Centre, it is you. And that is the reason I am here."

I realised the colossal effort it involved in addition to my heavy workload, especially since I had no other officer under me. I also realised that I had no choice. I told him to have the material sent to my office and leave it to me.

I was in it and there was no getting away. I decided, they will be the best training precis ever written in the Indian Army. Over the next nine months, I worked every day from nine at night to two in the morning on those precis. Five thousand copies of each were printed which disappeared like hot cakes.

I then wrote the fourth training precis on Drill on my own. That was the first drill precis ever written in the Indian Army, even before the IMA wrote one.

A lot of hard work, but it was fun, simply because I had the right attitude and brought myself to love what fell in my lap. This is the secret of a happy military journey.

A great attitude helps you live a stress-free life.

The Public Sector Units are white elephants the government continues to feed. I dealt with some during my service in the Mechanised Forces Directorate. The Ordnance Factory Board (OFB) was tasked to produce a practice anti-tank round to replace the expensive imported live rounds which are being used for field firing practice. This round was expected to be indigenous, reduce cost, reduce the barrel wear and tear, and have the same ballistics so that it could be fired from the same sights without adjustments. All very valid parameters. After years of so-called research and development, they just replaced the explosive charge with sand and produced one, which met none of the laid-down parameters.

Scores of meetings and briefings took place in their effort to get us to accept what they had rigged up. One day, as their project manager came to our office to discuss the progress on the practice round, he made a very humble request:

"Sir, do us a favour. Do not take this practice round; just say that we have made it so that we can put it on display as our achievement."

Well, many of these units work at sixty percent capacity and claim full overtime. God save this country.

The Regimental Days: Marching with the Fifth

I was commissioned in 5 GUARDS in June 1979. Well, in IMA, I was an optee for the GUARDS missile battalion. Despite being first in the order of merit, I was allotted 5 GUARDS, an infantry battalion, and another cadet commissioned into a missile battalion. Then Captain Lidder from Grenadiers (later Lt Gen), an instructor at the IMA asked me,

"Believe you are going to 5 GUARDS. I know you had opted for GUARDS missile battalion."

"Yes sir," I replied.

"So, are you asking for a change? You are entitled to get what you asked for. I do not know the missile battalions, but I have served with 5 GUARDS. All that I can tell you is, if you go to 5 GUARDS, you will never regret," he said.

I said, "Sir, you are from Grenadiers, you have nothing to take in the matter. But if you think that way about 5 GUARDS, that is the way it stays and I am going to 5 GUARDS then."

I joined the battalion and it was the wisest decision I made. It has been a story of two decades of wonderful soldiering, for 5 GUARDS was forever in field and operations and was rightly known as 'The Fighting Fifth'. Most of these years were spent in the most difficult of places in difficult environment. It was precisely these difficult times that brought out the best from these wonderful highlanders hailing from the lengths and breadths of the Himalayas from the J&K to the North-East. These wonderful soldiers never complained of anything and brought the battalion glory, time and again. These soldiers gave us great moments of pleasure through their simplicity, dedication, loyalty and above all their sense of humour.

It has been ages since I joined my battalion in 1979, and fairly long since I relinquished the command in 2002. I remember most of my stay very clearly, as if it was only yesterday. I stepped out of the train at Jalpaiguri to join my unit on a warm sultry day in July 1979. I was received by a well turned out senior subaltern, Lieutenant S.J. Singh. I climbed up in the rear of the vehicle, called 1 Ton, the Second World War petrol guzzler, anxious, wondering what lay ahead. Guardsman Harka Bahadur, a solid muscular Gurkha driver, slammed the tail board. The vehicle rolled off because of the impact, as the hand brake, while pulled up tight, was only ceremonial. Instantly, the thought crossed my mind, 'Wow, with such men, it has to be a solid battalion.' And it was. Harka Bahadur stopped the vehicle by brute force. The vintage 1 Ton refused to start as if protesting the punishment meted to its tail board by the driver. After the alternative starting mode of 'danda' start, a technology lost over the years, also failed, I was told to get down and push the vehicle. That was The Infantry, an arm driven on man and will power, thus never failing. That is how I push-started my regimental life. As we bumped along to the unit, an hour's drive away, I asked the jawan sitting with me in the vehicle, what he did for a living. He said he was the 'danda man' of the vehicle. Not wanting to expose my ignorance, I sat quiet, wondering what the 'danda man' was all about and what his duties in the service of the nation were. For me, coming fresh from the academy, every single soldier at all times was in the service of the motherland.

5 GUARDS was a 'pahari' (highlanders) paltan, with men drawn from across the length and breadth of Himalayas. The more time I spent in the battalion, the more I fell in love with these men. They were superb soldiers.

Less than a few months service, I had an encounter with the brigade commander. The commander visited my company during the annual inspection. Seeing the kit layout, he noticed that some men did not have identity discs. He asked me why? I replied they are young soldiers recently reported from the centre and have not been issued with identity discs. He said, "It is a simple thing, why are you waiting for it to be issued? Why can't you make it from a ghee tin?"

I said, sure sir, I could. And because it is so simple, why can't the country do so? I guess, I had announced my arrival.

The commanding officer visited my company in Sikkim. He entered one room and asked,

"Whose sahayak (buddy) stays here?"

I said, "No sahayak sir, I stay here."

"Why can't you stay in a bigger room?"

"The bigger room has been given to the men, we are short of accommodation," I replied.

That was the way I began, and that was the way I lived. The men always came first.

In the larger part of my unit life, we were out in the field and/or operations.

I have many memorable experiences that cannot be accommodated in any book. The life those days was different. The cross-border terrorism in Kashmir had not started. There was ample time for unit routine, PT, games, training et al. Inter-battalion competitions were contested with great zeal bordering on conflict at times.

Young officers spent most of their time learning the tricks of their trade, participating in inter-battalion games, learning regimental history etc. Thankfully, there was no cable TV, WhatsApp, Facebook etc. those days.

When I joined the battalion, most of the officers had participated in the 1971 Indo-Pak War. It was good learning to grow up with them. They were tough on you but knew where to look the other way.

Those were the days when there was no internet, no ATMs, no TVs and very few banks. The PNB, the mother of all banks where we all had our accounts, was twenty kilometres away in Siliguri. Dulabari on the Nepal border was not too far, from where people got imported goods, electronics being the craze those days. The consumer market those days in the country was basic to say the least and hence the craze for imported goods available across the border. Seeing others, I too decided to buy a "music system" as it was called. Hardly a system, it was a basic transistor sized Sanyo single cassette player. Those days it was the only entertainment available to a young officer. It actually cost a fortune, a whopping 900 bucks when my basic salary was Rs 750.

The bank being too far, and since I owned no transport, getting to the bank was like going on a pilgrimage. I sent my buddy to borrow Rs 400 from the unit baniya. The buddy returned empty-handed, saying the baniya has refused to give money unless I gave him a bank cheque first. The mess was four kilometres away and I needed the money urgently. That was the first time I was taking money from the unit baniya, which was a regular practice because of the distance to the bank. The officers would give him a cheque which the baniya would encash and square up. That hurt my pride. I was not running away from the unit for 400 bucks.

As the luck would have it, I was the duty officer. I ordered the weekly firing fighting practice in his baniya shop. All the jalebis, burfis and samosas were declared unfit for consumption and thrown out of the window. The fire-fighting party was told that the shop is on fire. Two fire extinguishers were let loose across his shop in full glory, turning his shop into stalactite. The baniya in chief, unable to believe what was going on, kept protesting to no avail. He was declared a severe burn casualty and ordered to be evacuated to the MI Room. The men understood what was going on, wrapped him in a blanket, loaded him on a stretcher and took him round the unit on the chants of 'Ram Naam Satya Hai.'

I got the baniya back and told him that I would not run away with 400 bucks. Don't ever make this mistake again.

The next day I was summoned by the second-in-command, Major Tuffy Marwah, a hard task regimental officer.

"The baniya complained to me what you did to him and his shop yesterday. Is that true?" he asked.

"Yes sir," I replied.

"Why did you do that?"

I told him the story, emphasising that I would not run away with 400 bucks. That he lives off the battalion and he cannot insult an officer by refusing to loan 400 bucks for a day.

"Why didn't you tell me rather than take this drastic step? You think I would not have taken appropriate action?" he asked.

"It hurt my pride as an officer. I did what I thought must be done so he dare not insult another officer. It is not in my nature to go complaining, and sir, surely you would not have done what I did." I replied.

Major Marwah smiled and asked me to dismiss without a further word. He ensured that my boldness and aggression was not killed. The matter ended there and nothing was ever heard of it again. If the commanding officer ever learnt of it, he never let it be known to me. My unit thereafter had the most disciplined and responsive baniya, for the regimental folklore never die.

Deployed in Sikkim in 1979, I had my first unique experience as Mr Sengar, as the Second Lieutenants were known those days, during the Dussehra celebration in the paltan.

A hill tribes battalion, Dussehra was celebrated in a big way over two days. Past-midnight, each company would sacrifice one goat that was served as prasad with a tot of rum. My commanding officer (CO), Lieutenant Colonel Pup Mann was an excellent commander. He would make it a point to project his officers wherever opportunity presented. As the sacrifice ceremony was on and the goats were being sent to their heavenly abode, one at a time, the CO directed that I do the honour and slaughter one. Never to say no, I was there with a khukri in my hand. The Subedar Major walked up to me and cautioned me to be careful and not to hurt my legs in the process. That was a good advice as I had never handled a khukri before, let alone slaughter a goat. His caution made me rather cautious. The goat was dragged to the altar, after kind of forcing its acceptance to be sacrificed by pouring half a bucket of cold water. I stood there with a white cloth tied around my head like a samurai warrior. As I swung the khukri through the goat's neck, the goat struggled and moved. The khukri slashed through its neck, leaving a sliver of flesh connecting the head to the body as it slumped to the ground. I struck again to complete the job. Well, I had failed in the task to chop off the neck in one go; never mind the margin of failure. I was penalised by the CO—a peg of Hercules rum to the whole battalion. It cost me better part of my month's pay. Great memories.

Four decades later, I can recount the whole evening to the last detail.

In those days, that seem so distant, the Hurricane lantern was our companion for the most part when the sun parted company every day. Petromax, the luxury of the days, was expensive at Rs 400. It would need a Harvard-returned buddy, called a batman those days, to have its mantle last more than a day. The pumping technique was complex. It needed the steady hand of a surgeon to give the mantle a lease of life beyond one day and very few mastered the art. Net result: mantle not being available easily in those remote areas, and costly at one rupee a day, petromaxes were packed and forgotten, only to be taken out on a marriage anniversary. Generators were still non-existent in the battalions and inverters not heard of in the Army. So, there was little to do in the high altitude once the sun went down.

The unit life was great and there was so much fun with those men. The biggest killer in high altitude for an officer is boredom and loneliness, especially when you are the lone officer on the post. I spent a lot of my time with men, learning about them, their culture and sharing all they did on the post. Once, during a roll call, I asked them a question,

"What came first, the hen or the egg?"

There was a long silence as they whispered amongst themselves. Then one Pahari came up with a comment,

"*Saab, iske liye toh dimag larana parega.*" (Will have to apply brain to answer this question.)

I said "*Aap dimag larao, phir kal jawab dena.*" (Alright, you apply your brains, we will discuss tomorrow.)

We forgot about this episode. A few weeks later I walked to the company langar to have an impromptu dinner with the men. Quietly, I took position at the serving point and started serving the men. In the poor light of lanterns no one noticed my presence.

And these men were discussing the million dollar question, "*Yaar, lieutenant saab ne bada jabardast sawal poocha. Kisi ko bhi jawab pata nahin. Mein teen hafte se roj sochta hoon, pahle anda aya ya murgi ayi.*" (Yaar, Lieutenant saab asked a great question about the chicken and the egg. Last three weeks I have been struggling with it and wondering what came first.)

I patted him on the shoulder, "*Teen hafte se mein bhi yahi soch raha hoon aur mujhe bhi iska jawab nahin mila.*" (I have also been wondering for three weeks and I too do not have an answer.)

My presence at the serving counter was very hilarious for them and we had a great laugh on the discussion of the chicken and the egg.

When on leave, we waited to get back to the unit; it was a home away from home. These wonderful men bore everything with a smile despite the fact that we were repeatedly in tough assignments. The battalion fought every war since raising, from the liberation of Goa to the Kargil war with distinction. In extreme conditions high in the mountains, they would give you the first hot cup of tea, brewed over a smoky fire of wet twigs. Served with smoke blackened hands and watering eyes, it carried the fragrance of their devotion and love that refused to wither with time.

May 1983, Brigade Exercise: The commander of 163 Infantry Brigade, a Dogra Regiment officer, ordered a 40 kilometres' inter-battalion endurance competition, immediately after the brigade had finished a 40 kilometres tactical exercise in searing heat. A mile stone on a road was given as the finishing point which would be manned by the brigade staff to take the report.

The exercise itself was a professional joke, asking the infantry battalions to exploit an objective 40 kilometres away after an attack on a water obstacle. My commanding officer's request to give the men a refill of water bottle was denied by the commander on the grounds that there was enough water en-route. In the searing heat on the lower Aravalli ranges near Hoshiarpur, there was not a drop of water. We came across a small cement tank full of muck, possibly made for wild animals. The men rushed to drink that crap. When I stopped them, my commanding officer asked me to let them drink. I said sir, this is no water, it is all muck, they will fall sick. He said that they will die of dehydration if they do not drink it.

The exercise, unrealistic as it was, was effectively over after this activity. The troops had been on man-pack rations for three days. The cook house lorries which would have brought hot meal after three days, which marked the end of the exercise, were declared destroyed by the commander. This order meant that the units would walk over 80 kilometres in twenty-four hours in peak summer without food and little water.

Through the night, the battalions pushed themselves to the limits, as happens in a competition, only to find no brigade staff at the designated finish point to take report, nor any sign of the commander. Finding no one, the battalions broke off.

In the summing up, I, then a young Captain of three years' service, to everyone's surprise, stood up to make my point.

I asked the commander, "Under the conditions in which certain activities were conducted in this exercise, I would like to ask four questions.

One, given the capability of infantry battalions and its limitations, is it a viable task to ask the units to exploit an objective 40 kilometres in depth or you had a different purpose under this garb?

Two, you ordered the competition, which meant that the brigade force marched 80 kilometres in twenty-four hours without food or water in this heat. Do you expect the men to operate effectively this way?

Three, having ordered an inter-battalion endurance competition, how do you explain that your staff was sleeping and missing when the battalions reached the finish point at five in the morning?

Four, and more important, under the circumstances, was the commander not expected to show up and meet the troops at the finish point, or somewhere during the 80 kilometres route march rather than be in his caravan?"

The commander, visibly shaken, attempted lame justification, and questioned my physical and mental robustness. He asked me, "Have I answered your questions?" to which I replied, " No sir. The issues that I have raised are very basic. They are on leadership, viability of the task and basics of soldiering. You have not answered any of my questions."

In private, he promised to fix this 'bastard'. As the saying goes, 'fortune favours the brave', I was posted out of the battalion. But, he did endorse in my confidential report that the officer needs to improve his sense of self-discipline. What an irony! A man who lacked the courage to accept that his staff made a mistake, for which he should have taken full responsibility and held his staff accountable, was telling me to be self-disciplined, because I had the courage to point out major omissions in the art of command and leadership. That is the irony, the senior can write anything and get away. In a 360-degree assessment, I would have had a view on his leadership abilities.

The same commander in another incident that involved me and him, not knowing the complete facts, asked me to apologise through my commanding officer. My commanding officer sent me to the brigade staff officer, Major Datar, SIGNALS (later Lt Gen), who asked me to walk up to the commander to apologise. I explained to him the complete story and asked him "Who should be apologizing, me or the commander?" I said that a junior does not have to apologise just because he happens to be a junior. If at all, the man who erred must apologise. If the commander decides to apologise to me, then call me, and I left his office.

Lessons

The commander, while having good intentions of pushing the units to the limit in an exercise, lacked the sense of how to do it. He was incompetent to task the brigade correctly. He fed false information to the units on the important issue of water en-route and by denying the battalions a refill of water bottle displayed lack of basic sense of right or wrong. His foolishness put soldiers to unwarranted risk in the searing heat.

While it may have been fine to deny the men a meal as an endurance exercise, he should have applied the same to his staff and himself. They were all on exercise and if the troops were asked to go without food and water, they should have done the same.

Most important, he lacked the basic understanding of command and leadership. He was a commander appointed by Army orders, but he lacked the basics of leadership.

More importantly, commanders of character do not give excuses, they accept their mistakes. It only makes them taller as a commander.

In the unit, we address each other with respect as 'aap'. If the commanding officer had to punish a soldier, which was rare, he would say, '*Aap ne galti kiya hai, iske lie mein aap ko quarter guard saza sunato hoon,*' There were no raising of voices and no cursing. It was a cheerful happy battalion. Officers and men greeted each other with a smile and with a resounding 'Ram Ram saab' from miles. The taste of langar chai and dal soup still lingers. The melodious sound of the pipe band and the lilting notes of the jazz band, practicing under a shady tree, still ring in the ears. The noisy training area was a happening place, with the mortar platoon being the noisiest. The battalion bara khaana was a cultural journey across the Himalayas, from the J&K to the North-East. A Kashmiri dancing the Bihoo and a Naga doing the Kumaoni dance were an utter delight. The officers and men mingled and it was an occasion for the men to let their hairdown. They were wonderful simple soldiers bringing special skills and culture from all corners of Himalayas.

When we were in the valley combating insurgency, there were no bullet-proof jackets for the first few years. In 1993 the first lot came and all battalions were issued a few for the officers. The officers declined to wear them because the men were right up with them. The men too declined to wear them for the same reason. Funny or stupid, no one wore them and they lay in the store till we had enough for both officers and men to wear. That was the wonderful spirit.

I had that superb buddy, Naik Mahesh. He would shield me bodily whenever we came under fire. I said, "You do that all the time, with no bulletproof jacket. You too have two little girls, like I have two little boys."

He said, "I am your buddy. I am responsible for you. I will never be able to live in peace if something ever happens to you without the bullet going through me." Not many are fortunate to have such top rate soldiers. Creditably, when the battalion was to move out from Sopore, the hottest insurgency spot in the Valley, where it had been deployed for long, a delegation of locals appealed to the Governor of J&K that this 'chiddy fauz' (Garuda emblem) should remain there and we did.

Counter insurgency operations have their own challenges. The environment in the Valley has turned on its head since the time we were there in early nineties. You could literally get away with the 'proverbial murder' those days.

We had our challenges too. Operating in and around Sopore, villagers particularly the women came out in large numbers protesting violently every time we apprehended a terrorist, shouting '*bekasoor-bekasoor*' (innocent). We tried to circumvent the problem by covering the apprehended terrorist with a water proof or blanket. It did not help as the protests continued.

Captain Niranjan of my battalion, known as Qayoom Khan, a wonderful soldier, said, "I will sort them out next time."

He came up with a novel idea. The next time Qayoom apprehended a terrorist, he covered him with a blanket and brought him to the centre of the village mercilessly kicking and thrashing him. The villagers raised hell shouting 'be kasoor-be kasoor, allah-allah'.

Unperturbed, he continued to thrash him relentlessly as the women became restless and violent.

He then asked the villagers, "How do you know he is be-kasoor?"

The villagers said he lives in this village and they know he is be-kasoor.

Qayoom then quietly pulled out the blanket exposing the man under it. It was one of our soldiers.

The villagers were in for a rude shock and sheepishly dispersed. We never had a problem again in that village.

Complex problems sometime have simple solutions.

Hav Rajinder Neog, an outstanding musician, suffered long from blood cancer when I was commanding. A little before he died, he said in English, 'Saab, 5 GUARDS is great.' That was the sense of ownership in all ranks. There was that simpleton and illiterate Subedar Major Dewan Singh who declined to proceed on leave on his daughter's death, because there were soldiers' cremations to be done in the unit the next day. Rising to this rank purely on loyalty, honesty and dedication, you could trust your last ten bucks on him. Who needs a graduate when you have such fine specimens as soldiers?

These men applied their own logic at times and that made life fun. Guardsman Kapson Pao, from Nagaland, returned three days late from leave. When he reported to the company Junior Commissioned Officer (JCO), he was told that he would be punished for coming late. Kapson Pao could not be traced the next day as the company looked for him everywhere. Pao reappeared after thirty days. When marched up and asked where he had disappeared after reporting to the unit, he said, '*Saab, main toh ghar ka problem se thora late aya. Senior JCO bola teen din late ho ya tees din, saza barabar milega. Toh main socha agar saza barabar hee milega toh poora tees din late aatta hoon, aur wapas ghar chala gaya. Ab tees din ke baad theek report kiya saab. Aap jo saza dena hain, de do.*' (I was late by three days due to domestic problem. Senior JCO told me that late was late. The punishment was the same for late from leave, whether three days or thirty days. So I thought if the punishment was the same, I decided to come thirty days late and went back home. You can give me any punishment you want.)

Havaldar Ram Bahadur Nepali was an outstanding soldier. When his retired father, also from the paltan, wrote to the commanding officer that his son had not come on leave since long, he plainly confessed that he had not gone home for nine months. When apprised that he had been on casual leave, he said, fifteen days is too less to go to Nepal, since his village is two days' walk from the road head. So, he has married twice, one wife he keeps at Delhi for casual leave and one in Nepal for annual leave.

These men were a different breed. They were not perfect but they were simply as good as it can get. They were always cheerful, absolutely honest, truthful, loyal, hardworking and dependable. In my over twenty-three years in the unit, there was not one case of theft. You were sure they will never let you down and they never ever complained of anything, and that made us more concerned for their well-being.

Out of twenty-three years, I served for about fifteen in the unit. There was no scope for a façade or double standards. We knew what the other was under the skin, and we were honest. We could recognise each other from the shadows and silhouette. The command of the battalion was absolute bliss. Young kids, who were toddlers when I joined the unit, were my young officers, with a large number of second generation men.

We were in the Siachen glacier in 1986. Those were tough early days. Our officers stayed in winters in the tattered snow tents and improvised shelters made from parachutes. The odd fibre glass huts that were there were given to the men. Not one officer was evacuated from the glacier on any ground. That was the culture of the unit. The commanding officer carried his own 'pithoo' and ran the BPET endurance run in battle order. We worked hard, laughed hard and drank hard. The tenure on the glacier with all its challenges was great regimental fun. The battalion loved being there and never lost its sense of humour.

Troops on the glacier were entirely air maintained. While air supply drops by Air Force transport planes brought hope and cheers, it was a risky affair. A wrong judgement by the pilot or strong winds drifted the parachutes dangerously close to the living area of troops. Besides, clearing the dropping zone (DZ) on a man pack basis was a real painful task at that altitude as all the snow vehicles were non-functional for almost our entire tenure. The only snow vehicle that worked was a snow scooter which only moved in reverse at a snail's pace, quite useless at best. A faulty drop by the aircraft meant an additional fifteen days' effort for the men.

One day the aircrafts dropped the load way off the DZ. One load fell on a snow vehicle parked way off, smashing the driver's cabin and narrowly missing the driver. Many loaded parachutes sailed over the living area sending the troops scurrying for their lives. Since you don't have anything worthwhile to discuss at that place, you discuss any event that occurs. I narrated the incident to Major Sushil Negi, the company commander at the Bilafond La post, located in one remote part of the glacier. His advice:

> *"Our Air Force is the biggest threat to own troops on the glacier. The only way you can save the snow vehicles from them is by using them to mark the DZ, right in the centre. That is where they will be the safest."*

I tend to agree with him. Ever seen an army four-ball on the golf course? All four aiming at the centre of the fairway and none getting anywhere near it, however broad the fairway.

Siachen Glacier is a brutal battlefield. One bad day can mean many lives lost. It is also a battlefield that separates the wheat from the chaff. The best memory I have from that challenging tenure is a lesson in leadership that a barely educated humble JCO from a remote village in Kumaon taught us.

Subedar Major Dewan Singh was the battalion subedar major. He was a God fearing simple highlander with a high sense of duty and loyalty. Always seen as not educated enough to shoulder higher responsibilities, he climbed up the ranks simply due to his loyalty, honesty, sense of duty and commitment.

One day we lost three soldiers to enemy artillery fire. The commanding officer was en-route to that post to meet the men. I was the adjutant at the battalion headquarters at Kumar Post, the central point of the glacier.

The same day a telegram was received that Dewan Singh's daughter had died of stove burst in his village. Very awkwardly, I broke that tragic news to him and said that he could leave at the earliest and I would arrange for a helicopter to take him to the road head, which otherwise was five days walk on the glacier.

What transpired over the next few minutes was humbling.

Dewan Singh after remaining silent for a while, said in Hindi, "Saab, this telegram must have taken many days to reach here. Even if I leave today, I will not be able to see the face of my daughter as the cremation must have been done long back. Today, we have lost three soldiers to enemy fire. Their cremation has to be done. I am the battalion subedar major. It is my duty to get that done properly and be present. I will only go after the cremation is over."

I saluted that man in my heart. A humble man who had hardly any formal education, was the wisest man I had seen, and giving me the finest lesson on soldiering.

It took us more than a day to transport the mortal remains to Kumar on sledges over crevasses, as no helicopters were available.

Those days soldiers' mortal remains were not taken to their families. They were cremated at the base camp. Unfortunately, because of the non-availability of helicopters, apparently due to VIP movement, the mortal remains of the soldiers could not be evacuated to the base camp. Firewood for cremation was air dropped by MI 17 helicopter at Kumar.

Everything burns slowly at that altitude. The wood finished but the bodies remained partially burnt. The bodies were then burnt for the better part of two days with ration cooking oil. This wonderful man, Dewan Singh stood there without food or water till the cremation was honourably complete. Only then did he proceed to deal with his grave personal tragedy.

And we are wanting graduate recruits. Soldiering is not about education. It is about passion, feelings, emotions, self-consciousness and brotherhood. This does not come with education. It is a result of good environment created by good leadership.

So long, Subedar Major Dewan saab. May God bless you, and the army be blessed with soldiers like you.

The battalion also launched the unknown attack to capture village Kirni in the Poonch sector in 1991, in less than eight hours after a stupid surprise order by the divisional commander, when deployed over forty kilometres in the mountains. After the attack, we held on to the defences dangerously dominated by the enemy from three sides.

In this horrible incident about which we will see more later, I saw the lack of professionalism at its worst: Generals and Brigadiers displaying a total lack of moral courage. I have invariably had issues with senior officers who did not live up to the standards expected of their rank and position.

In August 1991, after the capture of village Kirni, we remained deployed on the Line of Control in an eyeball-to-eyeball situation till the issue was formally resolved a few months later. We were dominated by the Pakis from three sides. Even eight feet deep communication trenches left us exposed from one or the other direction. While there was a fragile and uneasy ceasefire prevailing, the Pakis could hardly be trusted. The atmosphere was very tense and the troops were in a high state of alert. If they chose to attack, the Pakis would have been over us in a few minutes being no more than fifty to seventy yards, looking down on us from three sides.

Whilst the company was fully alert, there was a sense of uneasy calm with a suspect ceasefire in place. As a company commander, I would walk through the company defences very often. One day as I stood near a Paki bunker, a bulky Paki soldier came out. Seeing our slightly built Guardsman Bheja Ram, he stretched his arms and patted his thighs like a wrestler in a show of strength. Bheja Ram pointing at the Pakistani shouted "What are you trying to show me? Look at me." Saying this, he smartly did a hand spring and a couple of back flips on the bunker roof. It really impressed the Paki soldier. He shouted to his other colleagues who collected immediately on top of the bunker. The Paki soldier shouted, can you show that again? Seeing them gather, Bheja Ram shouted "*Ustad sirf ek bar namuna deta hai, chelon ab tum kar sakte ho toh kar ke dikhao.*" (Instructor gives the demonstration only once, my students, now do it if you can). The Paki soldiers watching the fun melted away faster than they had collected. So long, brave soldiers.

Half the company used to be on duty at night with lots of ammunition prepared ready to fire, and the balance slept in the bunkers fully loaded. I was sharing a big room with a dozen soldiers in a gujjar hut. While soldiers were all sleeping on the floor, my buddy from Meghalaya, Guardsman Noni Hazong had managed a 'look London, talk Tokyo' kind of a cot for me. Realising that I had been sleeping without a pillow, he asked me if I would like one. I said I won't mind if he could manage one. "But where will you find one here?" I asked. His reply was typical "*Saab, kuch na kuch toh karega.*" (Will do something or the other). True to his word, I was provided a makeshift pillow. It was hard and uneven. Under the circumstances, it was good enough I thought and complimented him on his resourcefulness.

A number of flag meetings later, it was mutually decided to fall back to the pre-crisis defence line. I directed that stocktaking of all stores and ammunition be done before we leave the post. The magazine NCO reported that all ammunition needed to be accounted for and he would like to lay them out to avoid problems later. Sure, go ahead I said, not realising what he was looking for.

"*Ja le aa*," (Go get them), he said to my wonderful buddy who was standing by, as if waiting for directions.

He disappeared and a little later strode out from my room with a heavy sand bag he barely managed to carry. Nonchalantly he started taking out 2 Inch Mortar bombs, laying them out—side by side in perfect discipline. Counting loudly in his typical Meghalya accent, one, two, tree...., he counted twenty one of them and said, "*Major ikees liya tha, ikees lotta raha hoon. Apka aur mera hissab kittab barabar, okay.*" (Sergeant, I had taken twenty one and I am returning them all. It's all clear between you and me, okay).

This fellow had come out from my room with a sand bag full of mortar bombs. I was wondering what the hell were these bombs doing inside my room.

I asked him, "Where have you got these from?"

"Got from your room, saab," he replied.

"And what were they doing in my room?"

"This was your pillow," he replied with a mischievous smile.

I was amused to see that all the bombs used for my pillow were without safety caps and half of them without even the safety pins. I had slept on that explosive pillow for over two months. And my meals brought from the company cook house were heated on a stove right next to my pillow. Since then I believe, if God is with you, no one can touch you, not even primed 2 Inch Mortar bombs in your pillow. I simply told him with a smile, with you as my buddy, I do not need the Pakis.

I love that guy Noni Hazong. He made sure I didn't have to worry about anything and that enabled me to do my job well in the tense environment. We speak regularly even today. People who criticise our buddy system fail to understand what that institution means.

Lessons

When you ask a soldier to improvise, you better check it out. It could give you more than a pain in the neck.

The environment on the post was very tense and we were vulnerable. Pakistan was smarting under massive loss of men and pride that they suffered at our hands. I had the experience not to ever trust the Pakis. They were dominating us from all sides and would have been over us in less than a minute if we were caught off-guard. Eight feet deep trenches were exposed from some direction.

Water had to be fetched from the way down; as such bathing was rationed or a man went down to the stream. The beard protected us from mosquitoes and fleas.

The Pakistani company commander conveyed a message that he would like to meet the Indian company commander. I conveyed to him to come over and have a cup of tea. He declined. I then said, if you are not coming, I am coming over. He was surprised. I crossed through the minefield hopping over the rocks.

I saw a very fair, smart, shaven Paki Major in a well starched sparkling uniform and shining boots with Rayban goggles. I wondered how he could live so ceremoniously dressed when his company was in a tense face-off. It looked as if he had not stepped out of an AC room for days.

When he looked at me, red-eyed and unshaven in a dusty combat uniform, he said in Urdu, "I wanted to meet your company commander and not a jawan."

I replied in English, "You are talking to the company commander."

He was surprised to see me and sized me from top to bottom, and said, "You don't look like the company commander."

I sized him up and said, "Neither do you look like a company commander of this company in combat."

He got my message and respect returned in his tone. He was surprised to see me hop over the minefield and walk into Pakistan. But I had my plans ready in case of mischief.

I signalled for tea. A soldier with a flask and a plate snaked his way through the minefield, walking over the rocks. Three steps from me, I watched that the soldier was walking straight in line with an exposed anti-personnel mine. My intuition told me his second step was going to be on the mine and I shouted for him to freeze where he was. Holding a plate with glasses, he was doing the balancing act and failed to stop. As I watched with horror, I saw in slow motion as he stepped on the mine and I waited for the bang. None came. I lunged forward to hold him where he was.

I said, "Don't move! Look down, you are standing on the mine." This wonderful soldier looked down and smiled, out of ignorance or relief, I will never know. I told him he was safe and asked him to lift his foot vertically and step on the rock.

A millimetre is no distance. But it was precisely just about that which saved his and my life or limbs. His boot heel was on the mine, but just missed the pressure plate by hair breadth. The curve of his heel, by divine grace, was aligned with the curve of the pressure plate and missed it by the least possible margin, resting on the body of the mine.

I have never been able to believe in the existence of God. I wonder who decided that this man should suffer no harm for my stupidity. I heaved a sigh of relief. Thank you God, if there is one. I would never have been able to pardon myself if that boy had suffered. I did not have to do what I did.

I told the Paki officer, these are the latest mines with a chip; they can differentiate between an Indian soldier and the enemy's.

Withdrawing from Kirni was some challenge. It was decided to flatten the gujjar huts manually without the Pakis detecting it, as it could have triggered interference by fire. It was some effort to bring them down with pick axes and shovels, all in one night. The next morning, the Pakis would have been shocked to see that the village did not exist anymore and the two companies gone, without they getting a whiff of it, less than fifty metres from them.

Later, deployed in the Valley, my battalion got caught up in a serious fire-fight with foreign terrorists in the built-up area of Sopore, in 1992. Sopore has always been a hot spot, even before the insurgency started. That was the time the place was teeming with hardened Afghan fighters. It was termed as the liberated zone by the people. All hell was let loose. The commander, a Dogra Regiment officer again, decided to get to the scene of action. The commanding officer, Colonel S.S. Kumar, a very competent officer, conveyed to him not to come there. He said, that the place is too dangerous and in that hell, he could not guarantee his safety and that the commander can influence nothing. On the contrary, his presence will become a liability.

In pursuit of glory, the commander insisted on going there. Once in the vicinity, he sent his staff officer, an education officer, to check out the situation. He reported that it was too dangerous to go, as there was fire from all directions. The commander derided him saying that he was a coward and sent him back again. The end result; this officer was pulled inside a room by the terrorists, shot and thrown on the road. Still not learning a lesson, the commander told his protection group to move forward and find a way. Soon one vehicle was hit and went up in flames killing one soldier. Another soldier who took cover under the vehicle was dragged out and shot. It took so much of loss to get some sense into the head of this moron commander, before he realised that the commanding officer was right.

I go back to the same point, how much are you prepared to suffer to give an incompetent infantry officer a compensation for his field service, even though the pro-rata was not in vogue then?

This commander had many other vices, which are better left unsaid. All this was known to the higher commanders. Yet this man was rewarded, sent on a foreign course and went on to become a Lt General.

Gen Patton, a legendary commander, was pulled up for slapping a soldier, and here all this happens, and the commander keeps going up in ranks.

The commander lacked the understanding of his functions as a commander. A commander's function in such a situation is to be prepared to support and facilitate the unit in contact. No commander should interfere in the battle of his subordinates, especially when the subordinate is a highly competent commander, which our commanding officer was. The commander's pursuit of an award had an officer and soldiers killed and given moral victory to the terrorists.

I was a flamboyant and a fun-loving officer with a good fan following in the battalion and the station. I had no fear of ranks or authority. In a manner that generated a cavalier attitude and a disdain for incompetent senior authority.

With my devil-may-care attitude, I would never close the top button of the shirt, uniform or civvies, no matter who was coming or going. Our commander in Udaipur, Brig Kahlon, would have noticed my habit and my drooping moustaches. He told my commanding officer to convey his displeasure to me and have that fixed. Col T.S. Chauhan told me what the commander had conveyed, but also said, you have always lived this way and you do whatever you wish to. That was great from the commanding officer. On the same issue much later, another commanding officer gave me a performance counselling.

This commander was the chief guest in our mess during a function. First, he sent everyone scurrying, saying he will have fresh coconut water. Those days, there were no bottled coconut water and no one ever surprised the units by asking for such a rare drink.

He then walked to me.

"Did your commanding officer tell you what I had said?"

"Yes sir," I replied.

"So, you have neither corrected your moustaches and you insist on keeping your button open."

"Just a personal fad sir," I replied.

He stepped closer, buttoned up my shirt and walked away.

Many days later, during the prize distribution of the inter-battalion handball tournament, he saw me from close range. He said, "You still have not trimmed these moustaches?"

"They have always been this way, sir."

Before it got worse, I got posted out from the unit.

Fortune favours the brave, as they say. Hats off to Col Chauhan for giving me this liberty to stay the way I was. I know, it must have been awkward for him.

We converted to a mechanised battalion in 1994, after I had put in sixteen years of service as an infantryman. With that ended the adventurous infantry life.

I always believed that one does not have to look and be serious to do serious work. As the second-in-command, I was stuck with a commanding officer who felt threatened by my presence. In the past, the commanding officers had exploited my commitment and enthusiasm to get the best out of me, and I gave my work my heart. They accommodated my fun-loving nature and candour. It gave me the freedom to remain myself and I gave my best.

Slowly, the commanding officer began to sideline me. Once he scolded the mess waiter for serving me tea before serving him during a floatation training. The poor waiter had not seen the commanding officer standing at a side. He would often countermand his own policies to overrule my directions.

Eventually, we came to head. My first report by him was poor in the context of the environment, a flat eight points from top to bottom. The second was a disaster. Fourteen six points out of twenty odd boxes, two eights and balance seven. I had no inkling of this surgical strike, even though it was not lost on me that he was no well-wisher.

Interestingly, he told me very innocently to submit my Annual Confidential Report (ACR) form as the MS Branch had asked for the cut-off report for the promotion board from Major to Lt Col. I was not expecting a resounding report. But what happened was a surgical strike in the true sense and quite interesting. Three days after asking me for the ACR form, he gave me a ninety seven point performance counselling, covering twelve pages. That strike achieved total surprise.

The intention was to get me pleading. He had made an error of judgement, yet again. Clearly, the performance counselling was compiled over the year with very precise details of place, date and the time in minutes; mostly, the creation of his complexes.

One of the points was that I was a poor example to the troops and the officers and that my top button was always open. My reply was in keeping with what I have always been:

> *"You were never there, when the bullets were flying. I sacrificed my peacetime postings with family to be with the battalion in action. If you have the guts, get the battalion fall in. You stand and I will stand. Ask them, whose example they would love to follow-yours or mine. You will have the answer."*

The commanding officer rather than talking straight to me and devoting his time and energy on the battalion was wasting all that on sorting his deputy out. He lacked the courage to get the ACR signed by me in person and sent it to me by post within the unit. Funny!

Long story, the corps commander was manipulated exploiting his association. The corps commander, Lt General Sangra, Dogra Regiment again, my senior reviewing officer in chain, endorsed the following remarks on my report:

> *"The officer will prove dysfunctional to the organisation if promoted to higher ranks."*

He too did not send the extracts to me for signature and this was communicated by the MS Branch months later. I sought the interview of the corps commander. Unfortunately, by the time I was given the interview, the corps commander had changed and Lt General Shankar Prasad, took over. I withdrew my request stating that I was interested in the interview of the previous corps commander, and not this one. The divisional headquarter declined my request with this funny logic that the corps commander has been informed so you have to go.

I told the corps commander that I did not want his interview and why I was still there.

He said, "You have a better record of service than I had at your stage. You think you can survive by confronting me? I am not here to listen to you. I will speak and you will listen."

He gave me a moral lecture. I told him that if he was not giving me a hearing, this interview has no meaning.

A court of inquiry was ordered. Both the commanding officer and myself were reprimanded. I, for not reporting things on occurrence, and the commanding officer for misuse of authority. He found it difficult to serve thereafter and left the Army.

The commanding officer tried to get some officers to give statements in his favour. Full credit to them, they all spoke the truth.

Lessons

The issues at hand were minor. The egos had clashed. When egos come in the way, logic goes out of the window. When the commanding officer started disregarding his own directions to compromise my authority as the second-in-command, trouble could not be far away.

In the whole exercise, there were no winners. The unit lost almost one year at a critical time of conversion into a mechanised battalion as the commanding officer was focussed on the wrong thing. The commanding officer chose to leave the Army. I suffered in no significant way, except that the unit lost the value of my contribution. Possibly, this report kept me from being nominated for the Higher Command Course, a minor issue.

How is it that I was graded outstanding in operations by three preceding commanding officers and he found me unfit for promotion from Major to Lt Col in a peace station, as did the corps commander?

Are there officers more suited for operations than serving in peace? If the answer is yes, then are they likely to suffer in peace? Then does it mean that you may be promoting officers in peace who will or may not deliver in war?

What could I have done differently? Possibly, bottle up my personality, something that made me an asset with other commanding officers in operations, and adapt to the commanding officer's way of subdued living and thinking. A subordinate has to adapt to his superiors who change, on an average, every year, and they could be as different as apples and oranges. When I retired, my last twenty-three annual reports were written by twenty-two different initiating officers. While that is a challenge in itself, what would be left of the officer's personality? Is that necessity creating a breed of 'yes men' soldiers? And 'yes men' soldiers are no soldiers—they are the enemy within. What does adapting mean, when a commanding officer, as was mine, took note of the distance at which he was when one got up from the chair to wish him as he approached you or how much was the distance when you did not wish his wife in a gathering? It is bound to affect the officers' individuality as a soldier.

How different were the other commanding officers who reported me outstanding in operations consistently? One, they were highly competent officers and had no complexes with anyone. They had all served in wars, operations and field with the battalion. They were extroverts and not the scheming types, who shared what was in their hearts and had fun. They accepted me as I was, extrovert, flamboyant and fun-loving, may be blunt at times, but well meaning. They realised that my heart was in the battalion and that given a free hand, I could be blindly left alone and trusted. And that is what they did. With that trust and freedom, to me nothing mattered except the battalion. There were winners all around.

The commanding officers must never expect officers to behave like him. He must accept and respect everyone as they are, value and capitalise the differences and exploit their strengths to maximise the results. The battalion needs the diversity of personalities to succeed in war and peace. If they are inspiring leaders, the officers will imbibe their traits anyway. Whenever egos come in the way, trouble is not far behind.

My statutory complaint against this report was rejected and it stayed in my record. But the Army is fair, at least till Brigadier level. I got promoted to Lt Col and Col on time, meaning this report did not affect anything till then. I missed the nomination to Higher Command Course (HC) because of this report, as I learnt from reliable sources. I later missed the promotion board to Brigadier in this pro-rata and quantified system.

My battalion maintained a piggery because of its class composition, i.e. hill troops and a majority from the North-Eastern states. Pork was their preferred meat, not to mention that it was a profitable venture and provided the much-needed additional funds to run the unit.

So, there we were, a well maintained and a hygienic piggery with massive pinkish Yorkshires. As the second-in-command I was responsible for all such institutions. I knew that a female pig, named Hema, was pregnant and there was excitement in Havaldar Kishan Singh, the senior safaiwala who was responsible for the piggery. I would get a daily report on the developments.

One day, I was surprised to see Kishan come to my house in the evening, panting. That was quite unusual. He looked very happy and excited. So I asked him what was up.

He said, *"Saab, sub theek hai. Sri maan, main aapko khush khabari dene aya hoon kee suari 'Hema' ne dus bache diye hain. Ismen se paanch ladies and aur paanch gents hain. Ap ko bahut bahut mubarak ho saab, aap kee hee maharbani hai. Saab Hema aur bache theek hain. Maine unke dekh baal ka pura bandosbast kar diya hai. Aap kal wahan ayie mein apko detail mein brief karoonga."*

(By God's grace, all is well. I have come to give you good news. 'Hema' has delivered ten piglets. Heartiest congratulations, it is all your doing and all the credit goes to you. Out of ten piglets, five are ladies and five are gents. I have made all arrangements. I will brief you in detail when you come there tomorrow).

I was impressed by his enthusiasm, and was amused when he congratulated me on the birth of the piglets. I swear I had no hand in the process.

I reciprocated his compliments and congratulated him as well. In fact, he deserved all the credit for he had looked after half a dozen pigs with devotion.

I went to the piggery the next day to see the new arrivals. I could not see any, and then Havaldar Kishan took over the proceedings. He removed a hessian cloth and there were two rum khokas. Each contained five pink cute looking piglets, no bigger than mid-sized rats, wriggling with their eyes closed.

> *"Saab jaise maine kal bataya tha, total dus bache hain, jisme, paanch ladies aur paanch gents hain. Command aur control ko madye nazar rakhte hue maine in dus bachon ko do section mein banta hai. Ek se paanch tak Number 1 Section aur chhe se dus tak Number 2 Section. Kyonki suari let kar dood pillati hai, at a time sirf paanch bachhe dood pee sakte hain. Is liye maine donon groupon ko alag-alag time mein banta hai. Har group ko ek baar mein 15 minute ka time diya hai. Is liye kee chatak bachhe dobara number na le len, pahchaan aur command control ke lie maine Number 1 Section ke mathe par kala teeka lagaya hai. Pahchaan ke ilawa inko kisi ki nazar bhee nahin lagegi. Alawa is ke, inme se sabse kamzor bachon ko Number 1Section men rakha gaya hai, jinka dood peene ka number sabse pahle diya jata hai."*
>
> (As I told you, there are ten piglets, of which five are ladies and five are gents. For the purpose of command and control I have

> divided them into two sections. From one to five is Number 1 Section and from six to ten is Number 2 Section. Because the mother pig lies down on her side to feed the babies, only five tits of one side are exposed. I have therefore divided the time between two sections. Number 1 Section feeds first and each section gets fifteen minutes at a time. To ensure that the smarter ones do not bully the weaker ones, I have put all the weaklings in Number 1 Section. For the purpose of identification I have numbered them from one to ten, and marked all the members of Number 1 Section on the forehead).

He then carefully picked up Number 1 Section and lined each one with a tit of the mother pig. I noticed that apart from the 'kajal' on the small forehead, he had numbered them from one to ten on their small butts. I clearly understood why, they all looked the same.

I was thoroughly impressed by his foresight, management and execution of an otherwise mundane job. I complimented him for his work and he got his best compliment from God, all the ten piggies survived. He looked after them as if they were his children. He said, "Saab, I have only six months to retire, please do not kill them for pork till I leave the unit." I promised him that and kept that promise.

The four-legged Guardsman, Badshah Khan, a Great Dane mascot, became an integral part of the battalion and brought everyone lots of joy. He would run the endurance test (BPET), receive visitors with a handshake, stand on hind legs and accompany the battalion battle cry with a synchronised bark. He was part of the battalion PT display team. Gradually he rose to the rank of a Naik, a rank bestowed on him in the unit sainik sammelan. Men often joked how Badshah was senior to them.

I miss those wonderful days.

THE MISSING ART OF COMMAND

I know I could be a difficult subordinate sometimes. One of my early commanding officers had written to my father to counsel me. While no one could touch me in my performance, my disdain resulted from my absence of fear of the authority in the army.

Less than three months into my service, I was to brief the army commander on the trials of the combat uniform that was being introduced into the army.

Despite numerous rehearsals, eventually approved by the brigade commander, I created a goof-up during the briefing of the army commander.

The brigade commander was thoroughly annoyed. At the end, everything seemed to go well during my briefing to the army commander, but it was not conducted the way the commander had approved. For sure, I had goofed up.

A few days later, my commanding officer Colonel Pup Mann (later Lt General) dropped in at the mess. "How are you doing Mikel?" He asked, my pet name in the battalion. I said, great.

After a drink, he took out a scarf from his pocket and presented it to me.

I was delighted, as those days, the commanding officers were rarely seen and little heard.

The next day, I was called by the second-in-command, Major Tuffy Marwah.

"Why do you think the Commanding Officer presented you that scarf?"

"He must be happy with me sir, why else should the Commanding Officer be presenting the scarf to me?"

"Like bloody hell! What happened in that briefing to the army commander? You seem to have goofed it up and the brigade commander wrote a stinker to the Commanding Officer conveying his displeasure. He thought, the best way to admonish you was to present you the scarf. One day you will get to know and never forget for years."

Those were the days.

Another incident, another commanding officer and me in the centre of action. A face-off with a senior officer in the division headquarters at Gangtok. He reported all over the place including my brigade commander.

Back in the unit, the commanding officer was much annoyed. Having given me his peace of mind, I was taken by him to the brigade commander. I stood outside the staff officer's office. After a longish time, he came out and we went back without me being marched up to the brigade commander. On whose directions he took me to the commander, I do not know.

I wish to highlight the difference in the approaches of the two commanding officers. My goof-up in the first case was far more severe, relating to the army commander's visit; the second related to a lieutenant colonel. A commanding officer must protect his flock, and deal with them himself as appropriate. He won't earn their trust if he takes them to the brigade commander like a class monitor.

Where do you think my loyalty is expected to lie?

My battalion was deployed on the Siachen Glacier in the mid-eighties. The Divisional Commander (GOC) flew to the battalion headquarters. He spent the night with our commanding officer, Colonel Umang Seth, a competent upright commander. He directed the unit to capture an enemy post which was way below our positions. Our posts dominated the entire pass and the enemy locations were much lower. The choice of the objective was clearly unsound. The commanding officer explained that he knew that area well and understood the implications, emphasising that it would become extremely difficult to hold and sustain the post under the enemy's nose. The GOC tried again, this time cajoling my commanding officer that he wanted to give a birthday present to the army commander and that he will personally lead 5 GUARDS with a band if they did that. Well, Col Seth told him that the plan was clearly made at the headquarters by people who do not know the terrain. He proposed that the GOC visit the place with him, and then if he concluded that it was a viable objective, the battalion will capture it. Needless to say, the GOC never went there and the matter was dropped. Well, dropped only for a while. When my battalion got a new commanding officer on the glacier, the GOC was back again with the same agenda. The new commanding officer Col Ranjit Singh was well briefed by the outgoing commanding officer. He too was convinced of the futility of the operation and had the courage to stand by his conviction. I am certain that their moral courage saved the army many problems. It was a case of strong moral conviction by both the commanding officers. There was a grave risk that both could have suffered irreparably in their careers. Both were men of strong character and prevented a futile expedition and a military blunder.

Let us examine another incident that my battalion had to endure in a warlike situation. In 1991, the battalion as part of 192 Infantry Brigade, 8 Infantry Division, was deployed in a counter-infiltration role in the Poonch sector. This meant a second-tier deployment in company groups extended over thirty kilometres in mountainous and high altitude terrain. The Pakistan Army had occupied an Indian village Kirni on the Line of Control which was the responsibility of 19 PUNJAB. The young men were taken away to be indoctrinated as religious extremists. The villagers had fled to safe hinterland. All peaceful efforts to restore the situation failed. The cross-border situation worsened graduating to the exchange of heavy artillery fire. It was decided to attack and take back the village. 10 MAHAR, 192 Infantry Brigade, in the neighbouring sector was tasked to capture the village with adequate time to prepare for the attack. My battalion was not given any task of any kind nor the Commanding Officer asked to attend the orders.. A day before the attack, Commander 192 Infantry Brigade directed Commanding Officer 5 GUARDS to attend the orders as observer as his company was deployed in the location from where the attack would be launched. On 31 Aug 1991, the commander of the holding brigade, 25 Infantry Division, was giving orders to his O' group. A most bizarre thing happened during the orders. The commander was interrupted by a phone call from his divisional commander. He returned and announced that the attack would now be launched by 5 GUARDS and not 10 MAHAR.

The Commanding Officer of my unit, sitting there as an observer, was struck by a bolt. The first thought that crossed his mind was, what kind of Generals do we have in our army? He did not protest there but spoke to his own commander and projected his problems and the major equipment deficiencies. The commander said that his requirements will be provided in location. It was near mid-day and with the battalion stretched over thirty kilometres in mountain/high altitude terrain, concentrating two companies, organising them for attack from a counter-infiltration mind set to an assault mind set without prior warning and without minimum inescapable reconnaissance and preparation was an impossible task. Add to that the fact that the battalion was short of major weapons and communication equipment. The dynamics of a battalion going into attack—physical, material and psychological—is complex. Ordering the unit to launch the attack in this fashion was plain lunacy.

Notwithstanding the challenges, the battalion took the task in its stride. It somehow managed all that was needed. But, clearly ill-prepared in every way, the battalion launched the attack. The radio equipment pooled from other battalions never worked. The unit for most part was out of communication during the attack. Our first casualties occurred when the patrol base of the holding battalion, 19 PUNJAB, which was our firm base/release point for the attack, opened heavy fire on my battalion as it approached the patrol base. This, when our guides came from the same patrol base. Helped by the rain gods and good conduct of operations, the village was retaken. Dominated effectively from three sides from less than hundred yards, the battalion held on tenuously despite heavy fire and beat back counter-attacks shredding the enemy over the next forty-eight hours. In less than twenty-four hours, the battalion had travelled a long journey; from a sedate counter-infiltration role to an assault on the enemy, having been launched into attack in the most atrocious manner.

The divisional commander had sprung the most unprofessional surprise on the battalion. He needed to justify his decision and be held accountable. What were the compulsions that suddenly emerged that he changed the assaulting battalion during the orders? A battalion which had not even been asked to attend the orders. Whatever dictated this decision was never shared or known.

Moral Courage. The fact that the battalion succeeded without too many losses is a credit to the resilience and training of the battalion, but of no relevance here. Amazing, that lack of own casualties was cited as the reason for not recommending the unit for more awards or unit citation.

Here, the question is, should the Commanding Officer 5 GUARDS have declined to attack the same day and insisted that he gets at least twenty-four to forty-eight hours for the attack? It is my firm belief that he should have done that, given his circumstances.

In this case, the Commanding Officer should have asked permission to speak to the corps commander. This would be moral courage and not cowardice. There was nothing that indicated that the attack could not be postponed by another day or, if that was not possible, the battalion that was nominated and had prepared went into the attack. I am certain, the Commanding Officer would have been morally right to represent against that order.

What role did the two brigade commanders play in this fiasco? Neither the holding brigade commander, who was giving orders, nor the brigade commander of 5 GUARDS legislated the righteous cause for minimum preparatory time for the unit for attack. A case of despicable moral and professional bankruptcy! The brigade commander of 5 GUARDS was morally and professionally bound to ensure that his battalion is not forced into this dangerous situation. I also wonder where the corps commander was during this all-important operation in the corps zone. While it is right to let the divisional commander run the show, he should be knowing which battalion is attacking and the change must be known to him. If I were him, I would have spoken to the assaulting battalion commander to wish him good luck, and check if I could help in some way. If he had done so, he would have known what was happening and intervened appropriately.

Full credit to the Commanding Officer, 5 GUARDS, to lead his battalion to a successful attack in the most difficult circumstances. And later, having the courage to reflect in the after-action report, the truth of the shocking manner in which the battalion was shoved into the attack, despite pressures from higher headquarters to change the report. This was the reason why the battalion did not get the credit it deserved. For records, I had drafted the report approved by the commanding officer and forwarded it. Both brigade commanders went on to become Generals—so much for our promotion system. It may not have happened if we had a 360-degrees feedback system.

Might sound silly, but it was our considered opinion that this was a result of inter-formation jealousy and friction and an attempt to let the other formation and the unit down. Can't believe that this can happen.

Similar situations occurred again during the Kargil war when the battalions were rushed into operations ill-prepared in the most difficult environment. Commanding officers who had the courage to demand minimum preparatory time for the attack suffered at the hands of incompetent senior commanders. Of the officers commanding the battalions in the Kargil war, only one made it to a three star—Lt General K.H. Singh, who commanded 27 Rajput, in the Turtuk sector. The plight of commanding officers in OP PAWAN, Sri Lanka, was similar. We are happier promoting untested peacetime officers based on power point and communication skills and sack righteous commanding officers who either question a wrongful order or make mistakes in operations.

Senior military leadership is expected to show more wisdom and moral courage in war and peace. In war when there are serious issues involved, a commanding officer or a commander must stand up and be counted. There can be no better illustration than the dilemma faced by Brigadier John Dalvi, commander of the Tawang Brigade in the Indo-China War of 1962. All officers must read his book, *The Himalayan Blunder*, a fine example of professional leadership.

Command Assignments

A Good Military Commander

knows when to just give and when not to give a damn...
irrespective of who it is or what the consequences are.

—Author

BATTALION COMMAND

The Army remains the finest profession for many reasons. Bearing arms for the honour of your country and dying for it remains the foremost, where you do duty for the sake of it. Your motivation comes from values that do not make sense to the outside world. Where the ultimate word in your life is 'TRUST'. Where your unwritten commitment to your buddy and fellow soldiers are stronger than the marriage vows of a couple.

For me, command assignments at all levels have been passionate love affairs. How I wish they had never ended! Sadly, I have seen many officers who were waiting for their command to finish, as if it was a thorn in the side.

For me, every command assignment was fun.

I had the opportunity to experience the dynamics of command and leadership in varied challenging environments. The leadership style that I adopted as a company, battalion, brigade and the divisional commander extracted the best from my command voluntarily, making every challenge feel like fun.

The army order gives you the authority to command.
Command does not sow the seeds of loyalty and trust, leadership does.

The ultimate challenge for a commander at any level is to create 'Trust and Credibility' amongst his command in his character and competence. These two are the qualities that will generate loyalty and commitment. Creating trust and credibility is a function of character, integrity, truthfulness, consistency in behaviour, genuine concern for your men and your ability to place the interests of your men before your own.

In our context, officers serve in the same unit for decades, thus we are all naked. When six hundred pairs of eyes watch you twenty-four hours, you cannot fool them for long. Our men are a superb lot, they begin by trusting you till you prove otherwise. But, in the present-day environment, that trust has to be sustained by the quality of leadership, competence and concern for the men.

In no other command than the unit command, are you more expected to sacrifice your personal interest for the men and the organisation. A soldier has much higher expectations from his commanding officer than he has from the army chief. Thus, it is pretty simple, if you expect your men to trust their lives on your judgement, should you not be prepared to sacrifice your career for them and for the honour of the unit? Which of the two is the bigger sacrifice?

During a Holi celebration in the battalion, I had two drums filled with rum. I invited the men to a 'no holds barred drinking.' The Subedar Major cautioned me that these men will lose control and create trouble. I said, I know them well, they won't. Surprised at the incredible offer, they hesitated. As I filled my mug with rum and drank, the men filled theirs and drank to their fill amongst much fun and chatter. Drenched in and out, they went to the barracks and went to sleep. There was not one untoward incident.

During this episode Havaldar Bhupal Singh of the unit put his hands on my shoulder and said, "*Saab, aap ek hukam dow, paltan ke 99 percent jawan apke liye jaan dene ke liye taiyyar hain.*" (You have to just order, 99 percent men of the battalion will lay down their lives.)

I asked him, "*Kitni daroo pee hai*?" (How much rum have you had?)

He said, "*Mein daroo nahin peeta saab. Dil se paltan ki baat bole raha hoon.*" (I do not drink saab. I am speaking for the battalion from my heart.)

I told him, that I was not pleased by what I heard from him. Somewhere, I felt that I had failed that there were still one percent men, that is seven in my unit, whose confidence I had failed to win.

Every commander has to be fearless. Either way, it will show in your body language. You can only be fearless as a commander if you have a principled life, have integrity, serve selflessly and accept your present rank as the last rank.

These days, anonymous letters against righteous people are common place. I got one as the commanding officer, sent to me for comments by the divisional commander. My response:

I ordered a sainik sammelan of the battalion. I told the unit of the anonymous letter. Then, to their surprise, I tore that letter in front of them and said, the man who has written this letter is welcome to write to the division commander what I have done to this letter.

I said, I command the battalion from my heart and give it 200 percent and command it in the best interest of the battalion and the men. As the commanding officer, I place myself at the bottom of the ladder for everything. No one, not even the Chief will tell me how to command my battalion. I am the Commanding Officer 5 GUARDS, and no one else has its well-being at heart more than me.

I wrote to the divisional commander that I command my battalion as per regimental mottos and traditions. In that context, I do not see any reason to change anything in the orders that I have passed, that are reflected in the anonymous letter.

The divisional commander appreciated my stand.

I was fortunate to command the battalion during Op Vijay, Op Rakshak in J&K and Op Parakram in Jammu Sector. I never wanted it to end.

I commanded the battalion through a set of maxims that I had laid for myself. These were penned down for posterity after the command as expected of the outgoing commanding officer.

When I missed the board to Brigadier, posted as Col GS (Operations) at Ambala, the commanding officer of an armoured regiment walked into my office. He expressed his deep shock on the board results. He said, "We have heard so much about your battalion command, that we are trying to emulate you." This was not the first time that I was given this compliment. I decided to write my perspective on command and leadership, in the form of my second book, *Battalion Command: Dare to Lead*, released in Jul 2016.

My Concept of Command

Hello Colonel,

> *Good commanding officers derive inspiration from their history but look towards the future.*
> *Battle honours on the wall cannot win the next battle, nor can the Gods.*
> *The battle will have to be won by your men and your officers, led by you, under your watch.*

Heartiest congratulations on assuming the command of your battalion. You have assumed the most honourable and responsible assignment in the Army, an honour which only the very fortunate have the fortune to experience. Eight hundred men will look up to you for everything. For them the world ends with you. For them you are the God. You will have to prove to be worthy of their trust. When you do that, your men will follow you to their death. This is my concept of battalion command, for whatever it is worth. Apply it as it fits your scheme of things. I followed it in letter and spirit and derived immense satisfaction and enjoyment during my tenure in command.

The unit command indeed was great fun. This will be the best period of your professional life. You have a 'Sultanate' in which you can realise your dreams as a leader. How selfless you have been in command will be reflected in how long your legacy lasts after you are gone and how much your tenure has contributed to your battalion's professional growth. While in command, be guided by the following principles:

- Formulate an inclusive Vision that should be shared by your command.
- Battalion command is a payback time; an opportunity to give and not to take.
- As a Commanding Officer, I would be scared of nobody, not even God.
- I would command my battalion from the head but lead straight from my heart. I will place the interests of my battalion, officers and men above everything else. I, as the Commanding Officer will come last, always and every time.
- My battalion will be the best operationally trained battalion in the formation. Anything that impinges in achieving this will not be

accepted. On every count, I must leave the unit better than what I inherited.

- This is my last rank; I aspire no more.

Every commanding officer has quotations that adorn the walls of his office. Generally speaking, they reflect his beliefs and values. I had the following that have remained in the office of Commanding Officer 5 GUARDS ever since and have guided the battalion well.

When an officer entered the Tiger's office, he read this quote:

> *'Hello Colonel, Command is a payback time, an opportunity to give and not to take.'*

When the officer faced the commanding officer, he saw this one

> *'No sacrifice is too big for the honour and glory of the battalion'*

And when a person left the office, he was left with this thought on the pelmet of the exit door:

> *'A lot can be achieved if it does not bother you who gets the credit.'*

Secrets of a good command are simple:

- Command is all about passion; develop this passion for the battalion in all ranks.
- Be a Tiger. To be a tiger as a commanding officer, you will have to live like a monk; in true Spartan style, living frugally. Detach from all materialistic desires that commanding officers often tend to fall prey to. It is better to be known as the commanding officer of the best unit in the formation, rather than a commanding officer who has the swankiest vehicle in town.
- Command is the authority entrusted in you to discharge your duties. It does not sow the seeds of loyalty and commitment, Leadership does. Lead by personal example; there can be no different standards for the commanding officer in any aspect of unit life.
- Prepare for the next war and not the last one. Your most sacred duty is operational readiness of the Paltan. Do not allow anything to come in its way. Environmental factors must be overcome; they cannot be an excuse. You must be prepared as if you are going to war next week.
- Think Team, Think We. The battalion is all about team work. Develop a happy cohesive team. Exploit every man's strength; harping on his

weakness does not do your cause any good. They are your wards, empower them, only then can they contribute to the realisation of your vision. Delete the word 'I' from your vocabulary, it is all about 'We'.

- Be prepared to accept the sole responsibility for all unit failures; credit for success is of the rank and file.
- Think long term: Paltan is an ocean and everything is important. There is no aspect of the unit that can be neglected.
- Your most important asset is the human capital. Invest in officers' grooming; they are the future of the battalion. Junior leadership, i.e., JCOs/Havaldars are the key to fighting successful battle in any environment. Their all-round empowerment must remain the key focus. Give them responsibility, accountability and respect and they will grow.
- Have a large heart. Your biggest strength is in pardon and not in punishment. Do not get perturbed by small omissions. Allow junior leaders to make mistakes without becoming failures. I have not seen one man who does not respond to a pat on the back.
- Your job is to train leaders and not followers. Battalion needs leaders at all levels to succeed in battle. Battalion should be able to function as effectively with all seconds-in-command in charge. There is nothing that you can achieve alone.
- Follow the Directive Style of Command. Develop decisiveness, fearlessness, mental flexibility and perseverance in all junior leaders.
- When there is a serious issue that affects your battalion, have the moral courage to take it up with your superior.
- Understand the nuances and implications of ACRs for the officers and men. They must not suffer because of your ignorance of the environment.
- The commanding officer's wife and ladies of the unit have an important role to play. She must maintain a happy domestic front, help in building a happy team in the unit and attend to the well-being of all families. She must play a positive role in the proper grooming of young wives. The first lady must never be allowed to assume a 'larger than life' role in the organisation and must never interfere in the professional aspects of unit life.

Professionally, she must leave the husband alone; he will do well by himself.

Focus on the genuine welfare of men: help the men to work towards securing good future for their children through better education and give them quality time with their families.

Create an Enabling Working Environment: Smile is the most effective tool in the commanding officer's arsenal for creating good atmosphere in the unit. Develop one if you are not blessed with one, and use it as often as you can.

Command must be fun. 100 percent work and 100 percent fun should be the mantra. Differentiate between office and off-office time.

Be predictable and consistent in your style of command. By the time you finish your command, you may add a few wrinkles on your face. They should merely indicate where your smiles have been – Have fun and enjoy your command, it will be over before you realise.

There will be no better time in your career when you can make a direct impact on so many men.

In mind, body language and action, be a Tiger and leave the pug marks of your character and personality in the sands of your unit history.

'Oh Lord, why can't you turn the clock back in time.
So I can be a Battalion Commander, one more time.'

*

These quotations have found their way into many other battalions and I am told that the commanding officers have found them to be effective guidelines. In my battalion, they were inscribed in brass by one of my successors and stayed for almost eight years. Then came a commanding officer who, quite surprisingly, removed the one that reminded him that he was there to give and not to take. Any commanding officer with suspect moral fibre will be uncomfortable seeing this board staring him in the face every day. He indeed was in the dock for the same reasons.

The quotes returned on the walls with his departure and have been there since. Those entrusted with the sacred responsibility to command a battalion, would do well to remember that while the command rests with them considerable power and authority; it entails awesome responsibilities. To sustain these, one needs broad shoulders and a big heart.

What makes a battalion tick?

A commanding officer's first objective is to be continually prepared for war and, second, to create an environment in which every man finds personal meaning and fulfilment. Create an environment where everyone is treated with dignity and respect and empower them to meet professional and personal demands and challenges. It is the attainment of the second that will enable the first.

An organisation will achieve extraordinary results only when every member of the organisation surpasses his potential. I always believed that a battalion will only succeed in war when it can effectively function under the seconds-in-command at all levels. Thus, a commanding officer must empower his deputy and deputies at all levels to depute for their commanders. It will happen when men have confidence, not only in their leaders, but in themselves, for the outcome in battle is a synergistic sum total of what each individual brings to the table and how the officers weave together individuals into a team and teams into sub-units and sub-units into a unit.

I followed this at the expense of annoying my superior commanders.

I was commanding my battalion under 26 Infantry Division. The new divisional commander, Major General Mohan Pande took over command and planned a visit to my battalion. Based on his visit plan, I planned my leave a week later. A day before the visit, I was informed that due to certain reasons the divisional commander's visit has been postponed to the next week.

I said, "Fine, but I won't be here. He is welcome, my second-in-command will conduct him." The staff officer was surprised.

"This is his first visit, and you are saying that you won't be there."

"Yes, indeed, that is what I am saying."

"Where will you be?"

"I will be on leave. I had planned it based on his scheduled visit, which has now been postponed. You can reschedule the visit after I return, or my second-in-command will take care."

"Is there some emergency?"

"No, that is my leave plan and my battalion can run in my absence as well. The second-in-command is meant to depute for me in my absence, even in war. That is the way I have empowered them."

The visit was conducted in my absence without a problem, even though the divisional commander wouldn't have been amused for sure. Compare that with today, where the commanding officers hardly leave the unit, believing that the unit cannot function in their absence, and the no-error syndrome.

Everyone functions more efficiently in a happy environment. The belief that men work better under pressure is a fallacy. The pressure may get you at best good performance, but won't get you outstanding performance and that performance will drop sharply in your absence. Good battalions are not those who produce results in the presence of the commanding officer; good battalions are those who are nurtured on good ethos and culture that needs no supervision. The officers and men must miss their commanding officer or commanders when he is away and not rejoice. It is possible.

Thus, a smile and a cheerful demeanour is the best weapon to create an enabling environment. Accepting mistakes enables an organisation to grow at double the pace. Every organisation must have a lessons learnt board, where an individual feels good about sharing his mistakes without a worry of embarrassment.

Crisis Management

In our profession, the prospects of things going wrong are very high. The biggest mistake that officers in command at whatever level make is to rationalise, for the so-called good name of the battalion. I believe, it is never for the good name of the battalion, it is to save the commanding officer's own backside and career.

I always believed and practiced that the shortest way out of the crisis is to tell the truth and accept the consequences. The Bible says:

Truth sets you free

And the best part of truth is that it needs no rehearsals. It gets you what no wealth can ever buy; mental peace and self-pride.

My battalion was conducting field firing at the ranges. I was given a shocker early morning that a soldier had died, even before the firing had started for the day. The previous evening, a violent sandstorm had struck the range that blew everything that had stood there. The whole day, every firer was given a fifty round belt. The last firer was given the remaining fifty-one round belt. As the sandstorm struck in full fury, one gun got jammed due to electrical fault and it was the last round, the fifty-first. The firing was called off to manage the storm. A live round remained in the chamber which went unnoticed in that commotion, as they deposited fifty rounds. Early next morning, one soldier entered the BRDM (Infantry Fighting Vehicle) to fix the problem. A while later, another soldier climbed on top of the BRDM and started cleaning the barrel. Having fixed the problem, he pressed the trigger to check out the gun, unaware that a live round was in the chamber and a soldier was cleaning the combat vehicle outside.

A big bang and a soldier dead in a second, a round having pierced right through his chest from less than three feet. A proud battalion that it was, the officers felt it would show the battalion in poor light. So, what was the suggestion? I asked. We will spin a convincing story so that the battalion's professional reputation is not tarnished. All officers were convinced that this was the best course of action.

I made the following points:

- You will be asking the men to tell lies. When you do that, you weaken yourself as a commander and lose the moral authority to command. You will forfeit all authority to look them in the eye and act against any violations in future.
- To conceal one lie, you will be telling hundred lies and you cannot conceal hundred lies being told by twenty men. You would have given official sanction to lying in the battalion. A despicable trait of a soldier and a disastrous character of a unit.
- Overall, you will invite more shame on the unit than this incident has done.

And that will be worse than what has happened, for it weakens the fabric of the battalion. In short, you will cease to be Tigers and will spend the rest of your service with the shame of lacking the courage to stand up and tell the truth. With what face will you command these men and ask them to live a soldier's code?

I rang up the divisional commander and reported as follows:

> "I am calling to report a tragic news. I lost a soldier today morning in a range accident because of mistakes on our part. I will fix what has gone wrong in the unit. As far as the General Officer was concerned, the entire responsibility for the incident rests with me."

The divisional commander said that no one would feel the pain more than me as the commanding officer. He told me to ensure that the next of kin was not put to any inconvenience and all the dues were promptly given and that the mortal remains were handled appropriately.

The matter was over in no time. I can never forgive myself for the sad loss of life in peacetime and the terrible consequences for the family. But I did not have to lose a moment's sleep from the guilt of not having the moral courage to stand up and asking my officers and men to tell lies. Such guilt can be very painful and long lasting to a conscientious officer.

A court of inquest was ordered by the divisional headquarter. No one was tutored. Every man said what exactly had happened. The inquiry was over in one day. My act prevented my officers from falling from the grace of our fine men and weakening the battalion.

Now, who is at fault in this very live incident? The outstanding NCO, who was detailed to fix the gun was charged with culpable homicide not amounting to murder. He was awarded loss of three years seniority that effectively eliminated all his promotion prospects to JCO rank. A bright career brought to an end, he left prematurely after becoming a Havaldar. I met this NCO and his family in his village a decade after I had relinquished the command of my battalion. We discussed that incident. And since then, I have been troubled by this thought that under the circumstances in which the incident happened, was I right to punish this soldier? What wrong did he do? Or, was I only finding a scapegoat because we had lost a soldier and someone had to be punished, and he was the convenient one? My conscience now says, I was wrong to punish him. There were series of unusual incidents beyond anyone's control that led to this tragedy; destiny as some would say. Yes, theoretically, the ammunition NCO should have ensured that the last firer returned fifty-one fired cases and all his ammunition was properly accounted for at the end of the day. But his whole establishment had been blown all over by the storm and it took us more than a day to find all the fired cases in the sand. Yes, theoretically, the NCO fixing the fault should have ensured that there was no one in front of the barrel when he pressed the trigger. On the other hand, a soldier was dead, and can there be no one responsible for the grave loss in peacetime? The more I think of it, the more I am convinced that he was awarded an unjust punishment. How much could the circumstances be accepted as a legitimate reason for the lapses by the crew and the ammunition NCO? While lapses did take place, as explained above, under the circumstances, who should actually have been held responsible for the death of the soldier? I now believe, circumstances played a big role in what happened. On the hindsight, I should not have punished that excellent NCO in a manner that ended his career.

Compare this with another incident when I was commanding a sector in a counter-insurgency area in the valley.

One morning, one of my commanding officers called up to say that one soldier has been seriously wounded during an encounter with terrorists. He gave me more details of the encounter. My staff officer told me that he did not believe that the report given by the unit is correct. I said, I will go by the commanding officer's word, for I expect my commanding officers to speak the truth. Later it was proved that my staff officer was right.

When I asked the commanding officer how serious was the soldier, he said that the wound was serious but he would possibly make it. A helicopter was requisitioned for evacuation from the helipad at my sector headquarters. The casualty was accompanied by an officer and his buddy holding an IV bag in his hand. Seeing no doctor or nursing assistant with him, I asked the officer why there was no medical staff with the casualty. He told me that the encounter was continuing and the commanding officer had held back the medical officer there, just in case they have more casualties.

The casualty was evacuated but succumbed to his injuries during the twenty-minute flight to the hospital at Srinagar. In the business of combat, some soldiers will die, sometimes due to unavoidable reasons and sometime out of negligence. In this case, the soldier had not died in an encounter with the terrorists. That was a total lie told to me by the commanding officer. The soldier had died during the training for search operations close to the company base. The young company commander simulated a terrorist by dressing up a Sikh soldier like one, trying to escape. He had not briefed the platoon that this will be happening. Though he had ordered the magazines to be removed from the weapon, the magazines were kept by the soldiers in their pouches. When the platoon was tactically searching the heavily wooded nala, the scouts spotted the movement of this simulated terrorist. The leading scout, unaware that he was a simulated terrorist, loaded his weapon and fired a single shot. Murphy's law went into motion. The single shot fired from an extremely awkward position on a moving target got him in the neck. Compare this with the number of rounds expended to neutralise one terrorist. In principle, the young officer did everything right as I had directed them to train. Only, he had not briefed the men properly and that cost the life of a soldier.

The issue here is not related to the mistakes of the young officer. The issue is related to the commanding officer. He fully well knew how the casualty had taken place, and that in this instance there were no terrorists in the equation. The company was half-an-hour drive from the helipad at the sector headquarters. The flight to the nearest military hospital was another thirty minutes. In all, it would take the casualty about an hour to get to the military hospital. And in that critical state, every minute of expert medical aid meant the difference between the life and death of that precious soldier. And to deny that man that aid to cover up their follies and trail of lies, when the medical staff in the battalion was doing nothing, was a despicable unprofessional act. In pursuit of his ambitions, the commanding officer had denied that soldier an opportunity to live. That commanding officer called me up at night to see me personally. When he met me, he said that he had been telling me lies about the incident the whole day. I told him I knew it all along and by now everyone knows it too. I told him what I had to.

The next day the corps commander was visiting my sector headquarters. The divisional commander flew in earlier. I told him what had happened and apologised for the wrong report. He said, he had given the same report up the chain and who would now inform the corps commander? I replied that this responsibility is squarely mine and I shall do it. I told the corps commander when he arrived and took the responsibility for the false reporting without deflecting the blame on the commanding officer.

Personally, I found that very disappointing. In counter-terrorist operations, things go wrong very often. The units hide many things, but acts such as these, leading to loss of a soldier just to cover your follies, in the hope of the next rank, is downright obnoxious. In brief, that commanding officer's lack of moral courage cost that soldier his life and lifetime misery to his family. A commander of conscience should be perturbed all his life for such acts of moral cowardice. As for me, it broke the bond of trust with that unit commander. I verified anything of significance when it came from that unit in future.

Commanding officers should watch out for a false sense of the unit's image. I am sure, more than the unit's image, they try to protect their own careers on this pretext. Often, there will be a conflict of interest between following the path of truth and unit loyalty. Manipulation of truth, misrepresenting facts, falsifying achievements for the so-called unit's reputation are ethically wrong and unpardonable. Where it leads to the loss of any life, it is absolute sacrilege. Such actions do great harm to the reputation of the officer class amongst the troops. It grooms junior officers with wrong values. Such commanders can have no claim to the loyalty of their subordinates.

Doctrine of Disobedience

I have never been a proponent of blind obedience of orders and I never have since my days in school.

As the commanding officer, one is expected to have his own mind and the ability to analyse the given order. A commanding officer must be prepared to represent against illegitimate, illegal or amoral orders, and if necessary, disobey them. Orders that will bring a bad name to our army and uniform, if implemented. Orders that will expose your men to unnecessary risks when not warranted; orders that cannot be implemented in the given timeframe or orders that negatively affect your command, etc. This is easier said than done, for in our army, there are neither adequate safeguards nor much hope of finding support from one's superiors. The officer in most cases is likely to come out battle (moral courage)-scarred at the end of the day. There are many ifs and buts and it is hard to draw a line. It all depends upon the conviction of the man and the readiness to accept the consequences.

Protect your command from incompetent superiors

Before you say, 'Good moral lectures'—what did I do in my time? I always walked the talk and practiced what I preached. I never let my men suffer because of the incompetency of higher headquarters.

I would like to narrate my two experiences, one in peacetime and the other in operations, when I had to make hard choices against the higher commander's orders, both at the cost of my career.

Located at Jammu, I was having a blissful sleep one night. Around midnight, the Colonel GS came on the line. He asked me to send one company to secure the Vaishnomata shrine by seven in the morning, as a VIP was landing by helicopter the next day at eight.

"That cannot be done," I replied.

"Why not?" asked Colonel GS, taken aback.

"It is midnight. You have given me no early warning for this task. The time and space itself indicates that even if my company was to start right away, they will not be there before midday. My men are asleep, they need to be selected, organised, briefed, draw weapons and ammunition, for which the kote and magazine need to be opened in the middle of the night, ready the vehicles with fuel, etc. They will be out for two days in wilderness, and need to be self-contained. All this needs time, particularly when you are giving me this task at midnight without any notice. This cannot be done by seven in the morning," I said.

"Are you disobeying my orders?" asked Colonel GS.

"No I am not. I am telling you that your orders are unimplementable in the given time frame. I will be committing a bigger sin by accepting the unimplementable task and risking the life of the visiting VIP," I replied.

"I am sure you realise that when I pass orders, I do that on behalf of the divisional commander. Should I tell him what you are telling me?" retorted the Colonel GS.

"Most certainly, the issue does not change, never mind who ordered it," I said.

He then asked me to order the party to start preparing for the task in anticipation, while he sought clarification. I politely refused, stating that I saw no reason in creating a panic in the battalion, when the task itself was unacceptable. I told him that I will wait for his call, only then will I take further action.

The Colonel GS, possibly realising the futility of pushing me, said he will get back. Half an hour later, he was back on the line cancelling his last directions and asked me to stand down.

The moral is, the higher headquarters sometime cover up their follies by putting impossible time lines on the unit. The battalion commander must weigh the task, and if it is impractical, have the courage to stand up, rather than put his men to unwanted misery. I am told that the VIP, in this case the corps commander, was informed at midnight of my stand. He cancelled his plans. If the person who created the 'faux pas' of failing to inform the battalion in time paid for it, I couldn't care a damn. I won't put my men to misery for someone else's incompetence, when there was little at stake.

The path of righteousness and integrity will often be a lonely trek. The path of idealism will mostly bring frustration in this opportunistic world. Yet, this is the path of professional soldiers that will ensure that you command respect and trust of your officers and men, even long after your authority is gone.

God seemed to have picked me to test my moral courage, time and again. Or was it that I was doing things differently?

After the terrorist attack on the parliament, the Army was mobilised in Jan 2002 under the name of "Operation Parakram". I was commanding a specialised battalion (Recce and Support battalion) at this time.

Two things happened that brought me into conflict with the divisional commander. First, my state of equipment was poor. Despite my best efforts and lack of spares for the imported equipment, nothing changed. My high decibel protests resulted in the MGEME (General rank officer from the technical maintenance service) from Command Headquarters to visit my battalion to take stock. He seemed least interested. He told me,

"Why don't you try TATRA vehicle engine in the BRDM?"

I was surprised at this question. I said, "Sir, I thought you are here to solve my problem. You are asking me to do a thing I have no means, resource, technical skill or the authority to do. You should be doing that. Isn't that the reason you have come to my battalion?"

That was the end of the higher headquarters' effort to resolve my problems.

During my briefing I apprised the GOC of the state of equipment and said, from the equipment point of view, my battalion is not fit for war.

He expressed surprise saying what had I done to correct the situation? I told him that he should ask the staff of what efforts I had made and why they did not keep him informed.

He deputed the Commanding Officer (CO) EME battalion to fix the problem. I told the GOC that the CO EME cannot influence anything. I have my own full-fledged workshop of 130 men. All that I need are spares which are not forthcoming for the last two years, being obsolescent imported equipment.

No spares came. All that happened was that this CO started giving false reports of my equipment readiness to show his progress. Once the GOC mentioned to me that I did not know the state of my equipment and the CO EME knew better. I said, no one is more concerned about my equipment than me, for my men have to fight with them in this looming war; neither does anyone know my battalion better than I do. That the CO EME was giving him wrong reports.

It would not only be wrong and unethical but dangerous for my men, for me to give an inflated report of my equipment readiness, as the GOC will task me accordingly. I would appreciate that the GOC trust the commanding officer's word and ask me and not anyone else.

I then told the Colonel GS of my conversation with the GOC and that my battalion was out of bounds for the CO EME battalion for wrong reporting. Either he conveyed that to the GOC in a manner he thought appropriate or the next time I will tell the GOC myself. The CO EME never again came to my battalion but got a VSM later alright.

The possibility of war at that stage was high. The units were all geared up and on the starting blocks, awaiting orders. Recce and Support battalions have a unique organisation and role. There is a vast variety of equipment and at that time they were in poor state for the lack of spares and vintage.

The companies were decentralised to the brigades and deployed in platoon entities. The battalion was deployed over a frontage of one hundred and fifty kilometres. This meant that exercising control was very difficult and necessitated the full-time commitment and presence of the battalion commander and the company commanders.

Much against my protest, I and my second-in-command were employed full time to run the divisional surveillance centre in the divisional headquarters that kept me away from the unit for days. A few days later, the company commanders of my battalion, deployed in the forward zone, were ordered by the GOC to establish and run the brigade surveillance centre in the brigade headquarters, without my knowledge.

When my company commander Major Narinder Singh told me that his commander was directing him to establish and take over the brigade surveillance centre, I told him to disobey that order and that he will do no such thing, unless I approve. I directed him to tell the commander in the same words.

The commander spoke to me very agitated and said,

> *"Your company commander is disobeying my orders and refusing to take over the brigade surveillance centre."*

I replied, "My company commander is not disobeying your orders, he is obeying mine and I told him to tell you what he did."

He said, "He is under command me and I can employ him the way I want. He cannot disobey my orders, and these orders have the GOC's approval."

I told him, "All this is news to me. I am the commanding officer, and I have the direct responsibility and the accountability for their operational performance and well-being of my unit. My company commanders cannot be taken away from their command functions without my concurrence. If these are the GOC's orders, I am afraid I cannot implement them unless I discuss them with the GOC."

I was briefing the divisional commander every day on the developments in the corps zone and represented against the GOC's directions. I explained that given various operational and equipment related factors, and based on my experience in the divisional surveillance centre, I am convinced that my company commanders will not be in a position to do justice to their tasks at this critical time, if they were employed for time-consuming staff jobs in the brigade headquarters. The General hardly heard me and rejected the request summarily and directed that the company commanders will take over the brigade surveillance centre and find time to command their companies as well.

I chose not to argue at that time. I went back and deliberated over the issue. I decided that we were talking war and there was simply too much at stake. I had a responsibility to my unit and to my men. My battalion was indispensable to the success of the division, and my company commanders were indispensable for the success of my companies. The decision of the divisional commander was neither in the interest of the division nor my battalion. The issue, I was convinced, needed to be represented in writing. The same day I made a written representation to the divisional commander. The extracts of the representation are reproduced below:

> *"The GOC's directions have resulted in my unit becoming headless; all command elements having been pulled out for staff functions in the brigade and divisional headquarters. The unit has serious issues with serviceability of equipment and the serious nature of tasks of the unit. That the unit, deployed along the hundred and fifty kilometres forward zone in platoon detachments will fire the first shots of the battle and take the first casualties of the looming war. The unit has been effectively reduced to a staff officers' battalion. In the absence of command elements, the companies have been left to inexperienced lieutenants and JCOs to fight the war under. My tasks directly impact the ability of the division to accomplish its tasks. In the absence of the command elements, I regret my inability to accomplish the tasks given to me by the GOC. I owe it to my battalion, that in this warlike situation, I strongly request for re-consideration of GOC's directions and the order be rescinded. I and my company commanders be reverted to lead the unit in war. I will provide suitable officers to perform the tasks at the formation headquarters."*

The representation was handed over to the Colonel GS of the division, who was stunned by the tone and tenor of the letter. He saw this as suicidal, given the fact that the GOC had already overruled the request the same morning. He also mentioned that my confidential report was with the GOC. I replied that I would rather go home as a Colonel who had the balls to stand up to the unjust order, than be part of battalion history as the commanding officer who got his men killed in pursuit of the next rank. Victory in case of a war-meaning unit's professional reputation was more important than anything else, not to mention the foolish loss of men, bound to happen without the command elements.

I was summoned by the GOC.

"I have gone through your letter. I am disappointed that despite giving directions on this issue earlier, you still represent in writing," said the GOC.

"You are right sir, you gave the decision in the morning. I felt that I did not have the opportunity to present my case properly and arguing there would have been wrong. The fact that I have decided to represent shows how adversely, I feel, the decision is going to impact my battalion's ability to accomplish the mission the GOC has given me. We are talking war and not an exercise, and therefore, I felt it was my moral duty to put forth my point clearly as it has serious consequences for my battalion, in case the conflict breaks out," I replied.

"I have deliberated on your case. I have decided that you cannot be relieved from the task at the divisional headquarters, but your company commanders need not take over the tasks at the brigade headquarters," said the GOC.

I thanked the GOC. I was happy that I had the courage to place the interest of the battalion and the division above my own, at a time when my annual confidential report was to be initiated. With my company commanders back with the companies, I was confident of dealing with the enemy. As for my report, it was of no consequence under the circumstances.

Did it impact my unit? Yes, it did. Everyone's appraisal report was well below what they genuinely deserved. Do I regret my action? No, I felt bad for the few company commanders, who richly deserved to command the unit but never made it to colonel. At that stage, war with Pakistan looked imminent. If the war had broken out, my battalion would have suffered grievously in every way if I had not questioned the divisional commander's orders. I would have never been at peace with my conscience if my men had suffered due to my lack of moral courage.

> *'Tell the King that after the battle my head is at his disposal, but mean time I hope he will permit me to exercise it in his service!'*

General F.W. Seydlitz to Frederick the Great, quoted in Christopher Duffy, The Military Life of Frederick the Great, 1986.

The point is, that if you have a case that you are convinced about, make your point with due military courtesy. Most senior commanders will see your point if there is merit, unless egos come in the way. If they do not, accept the decision and implement it with full commitment. You would have done yourself justice and earned a reputation that will give you a reason to stand taller and look your men in the eye, forever. Better, you earn yourself a lasting peace of mind for having the courage to stand up in the interest of your men. There is no better feeling.

Attila the Hun was a young accomplished king who achieved great military successes. His military chief, a middle-aged hugely experienced man was rather curt and rudely frank in his advice to Attila. When Attila pointed this out to his chief, his General said, "This way I serve you better, my King."

A unit will sometimes get awkward tasks with unjustifiable constraints. In the given set of circumstances, do the best you can. There is no reason to fret and fume over it.

A seminar was being conducted in Jammu University on Kashmir. Many military and civil VIPs were to deliver talks. That was a very sensitive time in Jammu when many terrorist incidents had taken place. My unit was made responsible for security with many constraints. In that university, no student or cars were to be frisked. No internal roads of the university were to be closed. No restrictions on the movement of students were to be imposed. No soldier would carry a weapon and the like. It was a perfect recipe for disaster. My representations were overruled.

I then decided what was best under the circumstances, prepared well and executed it. I put my men through intense pistol handling and firing practice and posted them at important places with pistols tucked under the shirts. A QRT had thoroughly reconnoitred the site in civvies and prepared a response plan in the event of a crisis. They were moving around the area as a routine troop movement.

On the day of the seminar, I sat nonchalantly near the entrance reading a newspaper. The brigadier in charge walked up to me and said that he cannot see any signs of security arrangements and seeing me sitting without a worry, made him very worried indeed.

I told him that I was given a task to be executed with both hands tied behind the back. Over the last few days I have done what I thought should be done and my time to worry was over. If the terrorists are smarter than me today, it will be their day, and the army would be looking for a head to roll and that head was mine. No sweat; worry won't help. I also told him, against all orders, I have armed my men and he could either report it or keep quiet about it. He said, "You have not told me." That was fine with me.

The brigadier said he liked my attitude and left me alone—how much assured, only God would know. The point here is that for any important task, it is important to prepare well in time and execute it well. A hassled and worried commander infects his command with anxiety and uncertainty. He has to always look in control and that would be very contagious in his command as well. In battle, things will go wrong for sure. The test of character of a commanding officer is to accept his mistakes and take corrective action. This will ensure damage control without further embarrassment and loss of lives.

I commanded the battalion based on the caveats I had set for myself. My style occasionally brought me in conflict with higher commanders, but they respected me for my convictions. The command had its challenges, having commanded in Op Vijay, Op Rakshak and Op Parakram. Occasionally, the battalion got un-implementable orders, but they were dealt with appropriately. There was never 'yes' said when military wisdom said the reply should be 'no.' Our men made the regimental experience an absolute pleasure. The universal ever smiling faces are etched in memory. I was most fortunate to have the finest team of officers and men that one can ask for. These paharis were lovable troops and we were one big family. There was passion in everything we did. The battalion was all that mattered and it excelled on all counts.

It was extremely sad to part ways from the band of brothers after twenty-three years in the battalion.

To this day, not a day passes when I do not think of the battalion and those wonderful men. It is a pleasure to get a surprise call from the men who served with me over three decades back. Those memories will go down with me to the grave.

After battalion command, I was selected to go as the defence attaché from a very distinguished field of ten officers. The corps commander, whose interview I had taken was the Director General Infantry then. I went to him for briefing before departure. He said he had seen me somewhere. I told him about the interview as a corps commander.

He said, "Yes, yes. I now remember. I am happy to see that you are going as the defence attaché."

I said smilingly, "No sir, I think, more than happy, you must be surprised to see me going as the defence attaché."

He had a good laugh, saying no, he wasn't surprised but happy.

Command Audit. When you finish your command, how will you answer the following questions?

- Were you selfless in command, or you pushed a personal agenda of self-growth at the expense of the unit?
- Did you have the moral courage to stand up for your officers and men when there was a cause to?
- Did you choose the harder right rather than the easier wrong and led by personal example on all counts?
- Did you attain a trust in your command in your character and competence?
- Will your battalion remember you as a commanding officer who made a difference to the unit, its men and their families? And the most important—
- If the battalion had the choice to choose a commanding officer to go to war under—would they choose you?
- The most critical—If your authority was taken away, how would your battalion respond to you?

SECTOR COMMAND

Lolab, called the land of love and beauty, is a serene place, ruined by terrorism. Indira Gandhi had spent a few days in Lolab during her honeymoon. Sadly, that hut is now occupied by a company commander as part of the company post. The horse trek along the hills, so popular with tourists in the good old days, is a dreaded place where only the soldiers dare to tread.

It is a highly dangerous place because of its geography and dense forests that start within a few meters from the village fringes in the valley. That gives the terrorists perfect setting to escape in no time. In this terrain, who lives and who dies is just a matter of micro-seconds based on who spots whom first, for engagements mostly take place at ranges of few metres. Since the troops always move in larger teams of ten or more, they're invariably spotted first, and shot at first.

The biggest vulnerability came from the fact that any one moving out of his base or head quarter had to return the same way. There was no separate route. Every few hundred yards was a culvert and a village hugging the road and touching the forest. In between the inter-village gaps, the forest came right up to the road; a perfect setting for roadside bombs, ambushes and snipers.

We trained hard with a focus, and by the time I was relinquishing command, we had killed the highest number of terrorists in the Valley, without a scratch to a soldier in operations. Sadly, we lost two men in the company base, because of poor briefing during training and poor make-safe drills. In both instances, it took one un-aimed shot to find the small fatal area on the men wearing the BPJ and bullet-proof patka.

I am a strong believer of 'Think Out of the Box—Keep it Simple'. Our training was based on three things:

- Ten most commonly experienced contact situations with terrorists and effective ways to deal with them.
- Field Craft-Making use of cover a everyday habit.
- Fire Discipline and Effective Fire.

The men believed that I brought good luck with me. During the past eighteen months before my arrival, the sector had lost men in double figures. In the first few months, men who by all probability should have died, survived without a scratch. The first took a AK 47 burst from less than twenty yards. Two rounds got his bullet-proof jacket and one grazed his patka. Shaken for a while, he got back in action. The second, amazing luck. Fired from less than fifteen yards, the terrorist bullet hit the soldier right above the centre of the forehead, but went straight through the gap between the head and top of the helmet, grazing and discolouring his cloth patka along the flight path.

On another occasion a bullet proof Mahindra Jeep went off a steep road. Rolling over twice, it landed on the lower road. The driver started the vehicle and moved along. No one hurt. All that was required was to replace the broken wind shield. The sector had no vehicle accidents.

One commanding officer told me, "The men believe that you have brought safety and good luck. As long as you are there, everyone will be safe." Isn't that a great feeling?

It was so true. Sometime after I left, the commanding officer of 18 Rashtriya Rifles (RR), Col Neeraj Sood died to a terrorist bullet. I believe,

> *"Incoming own fire is always more accurate and lethal than the enemy fire.*
> *There is no such thing as friendly fire."*

We had great fun as we trained hard and conducted successful operations. Since our operations were always without own losses, it was fun to sit with the men over a tot of rum. We played hard, had impromptu sports competition that took no preparation or effort, be it rain, snow or mud. All this was based on the conviction that the Battle of Trafalgar was won on the playing fields of Eton. Given the dynamics of a RR battalion, it helped to build good bonds and integration in quick time and break the monotony and drudgery of relentless counter-terrorist operations.

I had four Special Forces (SF) teams, three RR Battalions and a Territorial Army (TA) Battalion. The standards of SF battalions varied a great deal. One team did not have a contact in ten months, let alone get a kill. 9 Para SF team stood out of these. However, while they were good at pre-planned operations, they weren't so comfortable with impromptu operations that needed immediate response and briefing on the move. The RR battalions were excellent at impromptu operations as they knew their area like the back of their hands and had taped their drills for such operations.

Amongst many achievements, my sector killed four Indian Mujahideen terrorist from Kerala. That revelation jolted the authorities. We were told not to divulge the information and the police was asked to handle it. Since the local police invariably accompany the army units in operations in a built-up area, the press got to know. I got a call from one of the editors asking me about it. I said we have killed four terrorists, they are being identified by the police, he could get in touch with them.

This for sure was breaking news. He said, "Commander Saab, you know it. It is your units who have killed them. That is a big news. You do me a favour, I do you a favour."

I told him," I love what I do and do it honestly and seriously, without fear or favour. If ever I am in trouble, I accept my mistakes and move on. I need no favours from you or anyone. Get that?" He dropped the phone.

A strange incident happened in 18 RR (Raj Rif). During an operation, a soldier went missing. He was part of a team on a forty-eight hour operations in the forests. The patrol took harbour for the night. In the morning, he removed his bullet-proof jacket and with his buddy went to answer nature's call, no more than fifteen yards away in that thick jungle. The buddy returned alone, assuming his buddy was taking more time. When he did not come back, a search started and he was nowhere to be found. The commanding officer informed me about midday.

He was an excellent soldier doing important things. His relief had reported and he would have left in four weeks' time. The commanding officer's assessment was that he had been kidnapped by terrorists from the ablution site.

"How many yards from the harbour?" I asked.

"No more than fifteen yards." He replied.

I said, not possible that at that distance it can happen without a trace and noise or alarm being picked by the team so close.

The corps commander asked me my assessment,

"He has deserted." I said.

"Why should he desert when he was to leave in four weeks?" he asked.

"My hunch is some ulterior motive to disappear for good to claim the benefits of MIA (Missing in Action), believed killed by terrorists." I said.

Both corps commander and the commanding officer were not convinced by my assessment. Both believed that terrorists may have just pulled him into a hide right there.

My hunch was bang on. That is what exactly had happened; he had staged his disappearance. Why did I think it was the case?

When the commanding officer briefed me, I was reminded of a similar incident that had happened in the North-East, involving a medical officer, engaged in an extramarital affair in the same station. This officer had apparently died in a car accident and his body was believed to have been washed away in the torrid river, never to be found. This officer was picked up from down south a few years later. His plan was to settle down with his lady love anonymously in an unknown land.

A few months ago, in the same unit, a soldier had died during training in weird circumstances. No inquiry was necessary in the operational area—only a brief situation report was required and that was the end of the case.

The unit's version did not make sense to me and I was convinced my assessment was more probable.

He was finally traced to his father-in-law's house. He had planned to disappear for good, in the hope that he will be declared missing, believed killed by the terrorists and his NOK will get the dues.

He was brought back to the unit and he said, his wife was harassing him for jewellery. Also, his father had asked for a contribution from him to construct a house and his wife was quarrelling over it. Harassed and frustrated, he thought this was the panacea for his financial trouble and he would live quietly somewhere for a few years till the case was forgotten.

What did he do?

In the thick forests of Lolab, he walked away with the AK 47, unnoticed, as his buddy sat for nature's call. The previous day, he had hidden a pathan salwar suit in the hollow of a tree somewhere. He went there, hid his weapon, changed to pathan suit and walked away. Took a taxi from Lolab to Kupwara and from there to Jammu.

The buddy system failed to detect this man's intention.

Commanders must be truthful without exception and protect their command where they deserve, no matter the issue and punish those who repeatedly fail to listen to the voice of counsel and bring disrepute to the army.

During an operation in my sector in the Valley, one soldier had fallen in a nala on a freezing winter morning. He picked up some papers lying around to light a fire to dry himself. A passing villager saw the bits of burnt papers which happened to be Quran pages. All hell was let loose. The committee formed by the villagers filed a FIR and deposited the burnt remnants in the police station. The commanding officer was told to meet the village elders and the maulvis and resolve the issue. Rather than going himself, he sent his company commander who was a very junior officer. The crisis snowballed. The commanding officer then informed me of the crisis. I arranged for the village maulvi and the police officer to visit the site and see for themselves.

They were convinced that the soldier was not at fault. I drove straight into a crowd of couple of thousand agitating villagers against all advice. I addressed the rampaging crowd along with the maulvi. The villagers demanded that the erring soldier be brought to the village and be punished. I made it clear to the crowd that the Army has a system to investigate and take action. Prima facie, the Quran papers used by the soldier to light fire were lying around on the road. They had no business to be there. The prerogative to deal with the man was the commander's and there was no question of the soldier being brought in front of the villagers. The village was told that the inquiry will be completed in two days and the decision taken on the soldier will be announced from the village masjid by the maulvi in three days. That the Army prides itself in upholding the respect of all religions. That in the protection group of the commander itself, there were men from all religions.

The maulvi also addressed and admonished the villagers for the state of the masjid and the tattered Quran which flew all over in the wind. He endorsed my view that the soldier was not at fault. There are good maulvis too, as this one was who spoke the truth despite his vulnerability.

The massive crisis, with great potential to flare up, was diffused in quick time. The villagers broke off after my talk. The soldier was acquitted, and the decision communicated to the villagers as promised.

The higher headquarters was worried that the incident should not be reported by the media. I realised, in the present-day environment, with sizeable hostile population, there was no way to prevent it from appearing in the media. I called the media personnel and apprised them of what had happened. The next day, the incident was reported by only one paper in an insignificant corner.

Lesson

If you do not tell the media your story, they will print the other's story. Full credit to the maulvi who stood by the truth, in spite of his vulnerability in doing so. Remember, Truth needs no rehearsals. You will speak the same thing no matter how often you have to repeat it. And that means a peace of mind.

It is surprising how senior commanders panic in crisis. Hilarious un-doable suggestions came from higher headquarters. I was asked to do weird things to diffuse the situation.

Finally, fed up with the wisdom of the higher headquarters, I told the divisional commander that I understand the directions he is getting and from where, but requested him to leave me alone. It was my sector, it was my problem and I will resolve it. In the worst case, I am the commander and I take full responsibility for the incident.

There is always a reason to laugh, no matter the circumstances. The divisional commander was very worried how the burnt remnants, no more than a few thumb-sized burnt pieces of the Quran pages, would be disposed. When he was told that they will be buried as per Muslim practice, he was concerned that the villagers might dig them out and reignite trouble. I assured him that no such thing would happen as the remnants would be buried within the company location. When he still remained concerned, I told him that water will be poured to ensure that nothing is left of them by the morning. Still worried, he said that's fine, but they must be watered every hour through the night. He was put at ease by saying that it will be done.

The Sector Command in Lolab was fun all the way. I commanded the sector from the tennis court, which also had a golf practice facility made from golf nets. A telephone line ensured I was available to anyone who wished to speak to me, units or higher headquarters. We trained hard, operated hard, played hard and had great fun too.

Our PT and games parade were energisers with no holds barred fun and laughter with officers and men.

The Camp Senior JCO was a fun-loving Gurkha, yet serious in his job. Every now and then we had impromptu officers, JCOs and troop competitions. He would always volunteer to become the referee, even if he had no idea of the game. And the commander's team always won. Commander could commit no foul. When the men shouted from outside, *"Saab, aap commander saab ki taraf bhai bandi kar raha hai,"* he would reply, *"dost, fauz mein aisa hi hota hai"* and the game would go on. Great fun.

Disciplining the Tennis Balls

We had constructed a tennis-cum-basketball court and would often play morning and evening. The other officers had started learning the game there. Soon they picked up the game well and looked forward to it every day. Colonel Sudeep, our camp commandant, bought top of the brand tennis racquets for all—Babolat, Head, Prince and Wilson. Seeing them, any one would believe we could compete in a grand slam, but I guess the similarity ended right there.

The sector CSD had Cosco tennis balls and we enjoyed playing with them. All of us being novices, we had no idea of which balls to use on a cemented court. Sudeep had thoroughly got engrossed in the game he had learnt at forty-eight. During his leave, he took tennis coaching lessons and brought three dozen Penn balls. These were low pressure soft balls that felt great on the cemented court. At the rate these balls were being hit for sixes, soon we ran out of them and were back to Cosco balls. It was only then we realised the difference. These balls felt like stone and gave most of us tennis elbows. Some officers went on a self-imposed rest and the tennis court was less crowded.

The CSD, somehow, had two hundred Cosco balls in store and no one was interested in them. I decided we needed to play them out or else the CSD would suffer a huge loss. One evening, Major Kuldeep, an ever cheerful Texan, was waiting for his turn to play. I saw him seriously studying the two types of balls, pressing them alternately. He sent a jawan somewhere who returned after a while.

Kuldeep then tossed a Cosco tennis ball and said, "Sir, try this, how does it feel?"

We knocked around with this ball for a while and wow, it felt great, just like the Penn balls. "How did you do that?" I asked. He held up a syringe saying he had removed some air from the ball with it. How wonderful; a stupid idea that was the perfect solution to the problem that suddenly made all the rotting Cosco balls relevant. As the news spread, all our tennis elbow casualties were back on the court. Thereon, the syringe became part and parcel of our tennis kit, not for a medical reason though. Soldiers who helped us during the game became masters of syringe treatment and did a great job every time a new set of Cosco balls were brought. Life was back on the court and the Penn balls were forgotten.

One day a new set of Cosco balls was brought and subjected to the syringe treatment by the soldier. As we played with the new balls, they hardly bounced. Far too much air had been taken out with the syringe. We tossed the balls out disappointed that three new balls had been ruined. We were back to our old bald Penn balls that had hardly any felt over it. A little later, the soldier tossed the same balls back and asked us to try. We knocked with those balls and the lousy bounce was gone. They bounced well and felt great. The secret was simple. The soldier had injected some air back into the balls with the same syringe. Would you believe how the soldiers can improvise?

As the monsoon came, our tennis was disrupted. The balls would get wet and heavy and broke the racquet gutting. In the heavy rains, there was really no way you could play. Col Vishal Chadha said we can play after light rains. I said, "The balls would get wet from the grass around the cemented court." He said he had made some arrangements. I wondered what that arrangement was. As we played, the balls kept getting wet from the grass around the court. But time and again dry balls appeared from somewhere.

"How many balls have you got here?" I asked.

"Four," he replied.

"How are these dry balls appearing all the time?" I enquired.

"Microwave oven," he replied coolly.

A soldier on duty would disappear with wet balls, bolting off to the nearby officers' mess and be back after a forty-five seconds microwave treatment.

This treatment, while drying the ball alright, made it even harder and bouncy and our tennis game was closer to volleyball, good enough for the place we were in and we enjoyed every bit of it.

There was just one thing that always raised my pulse rate. It was the signature Airtel tune of my mobile, every time it rang. It could either be a good news of terrorist killed or the terrible news of own casualties. God be praised, if there is one. There was never a bad news.

I did not believe in God's power to save a soldier's life. Even the terrorist has his God. I visited my sector headquarter mandir twice in my tenure, on taking over and on relinquishing command. I stopped the practice of battalions taking me to the company mandir and the pandit giving me prasad. To me, it unnecessarily seemed to spread an impression that we were in a dangerous place and spread anxiety amongst troops and we needed God's blessings to survive. God, if there is one, will come to your rescue only when you empower yourself to meet the challenges you face and that comes from purposeful training and effective leadership. Keep sitting on your haunches and God will be happy to see your butts shot off.

In combat there are times when the commanders need to make decisions that are very awkward and debatable in their correctness. That is why, command is an art and not a science. We were heavily committed in an operation close to the location of 18 RR. The commanding officer, late Col Neeraj Sood, said there was a fire incident in the unit and sought my permission to leave.

When he returned, he said it was a minor incident and all was under control. I forgot about it. The next day the divisional commander enquired about it, and I said, I have not been to the site, but the commanding officer tells me, it was a minor thing and all is well. He told me to please check it out as it had been reported in the papers.

When I enquired again, Col Sood repeated the same thing. I asked him a few questions and I knew he was hiding the truth.

I immediately visited his unit. Cornered, he told me the story. The unit was selling 12000 litres of surplus kerosene. Things went wrong, the kerosene dump caught fire and the civil petrol bowser being filled, was reduced to ashes. The truck driver died of burns in his effort to save the truck.

I asked the commanding officer,

"Why were you doing it?"

He said, "As you know, an ex-serviceman's innocent married son had died in an ambush in his unit's operation. The father had agreed to settle the case out of court for a sum of Rupees five lakhs compensation. I wanted to settle it with this money."

"What do you want me do?" I asked him.

He said, "I know, the gravity of the incident. But I will request you to save me."

I told him, you do not know how much Kerosene was being sold, you do not even know at what rate it was being sold. It is simply appalling. Your battalion seems to be in utter chaos.

Nevertheless, I said, I will give you the benefit of doubt because I am aware of the case.

I told the divisional commander, the incident is serious and I will handle it at my level.

He lived to fight another day, only to die later in an ambush after I had relinquished command. Well, lots of things happen in those environments and lots of things are covered. That is what makes command fun.

I was sad to leave a team that was so well trained and integrated. A team that had achieved so much, smiling all the way, and raised my stocks in the Northern Command. Have never seen officers cry the way I saw on my farewell parade before I finally drove off. God Bless Lolab Tigers—be they always be safe.

MS Branch and its Amusing Ways

As I was nearing the end of my tenure, I was asked for my choice of postings. I wrote a letter to the MS requesting for a posting to New Delhi, where I had never been posted ever. I was told by the dealing officer (Director MS X) that there was no problem, as I had never been posted to New Delhi.

One day, the Colonel GS informed me that my posting order has been issued, but there is something wrong with the posting order. It says posted as Deputy GOC, 56 Infantry Division, but there is no such division in the army. I found out, it was a new division to be raised in Arunachal. So much for the MS Branch. From Lolab, Kupwara to Arunachal. My request for an interview elicited no response. I accepted it in stride and joined my new assignment, when a number of officers had gone from New Delhi for command of brigades in peace and went back to New Delhi. In the MS Branch, they told me that

I was being handpicked for this important job. Well, I was non-empanelled for Brigadier in the first board and now they were handpicking me as a special talent to raise a division! They surely have a method in madness which only they understand. Simply, I was not street smart enough to find out what was being planned for me and get it cancelled in time, like the rest.

Later, when I asked the Director MS X, what happened to my case, he said that the MS was hell bent on ensuring that I do not come to New Delhi, the MS being Lt General Awadesh at the time, of the Sukma scam fame. If what he said was true, I had never even seen this gentleman, let alone serve under him. Quite interestingly, later I discussed this case with a retired General who was the Additional MS at that time. He said, it was not the MS, but someone higher who did not want me in New Delhi, a name he did not divulge. The man above the MS is the Chief. I did not know the Chief either. Wonder what goes on!

One year later, when the division was raised and assumed operational responsibility, I wrote to the MS Branch, stating the past experience and that whenever they chose to post me, please do see the merit of my posting to New Delhi, against a dozen others who have been posted from New Delhi to peace for command and back to New Delhi.

I was on leave. To my surprise, the Additional MS found out my mobile number, expressed his regret and confirmed the receipt of my letter and that the needful will be done. I had barely reached back after leave that I was posted to the NCC Delhi Directorate forthwith without relief. From Arunachal to NCC without relief; some priority. Wow. I did not even know, where and what this appointment was.

When I called up the DDG NCC, New Delhi, the incumbent did not even know that I was posted in his place. He was then hurriedly attached for six weeks as additional officer. Thereafter, suddenly he was posted forthwith to an area headquarter in the south to be additional officer there till further orders. They certainly have a sense of humour and some sense of urgency.

Someone rightly describes the MS Branch, 'Show me the man, I will show you the rule.'

I was part of the pioneers who raised a new division in Arunachal Pradesh under (then) Maj General R.N. Singh, as the divisional commander. Not one man was attached in the divisional headquarter. The divisional camp managed the whole show including the Flag Staff House guard. There was much resentment because the officers were used to the comfort of excess manpower drawn from the units, who either can't or choose not to represent. In case of war most of these attached men will be returned to units, and in all probability launched into battle without current training.

As the Deputy GOC, I had a staff car and a light vehicle. In that field station, I rarely used a vehicle. When I did, it was never more than one vehicle at a time, that too for a very brief duration. I called for the staff car driver and asked him what he had done in the last four weeks. He replied, he waits in the division MT (vehicle) park the whole day for orders. That man had simply done nothing but wait the whole month. I directed that one of the two drivers be dismissed and only one driver will manage both vehicles. When required, the second driver will be called for from the unit. This suggestion surprised the staff officers, who gave all kinds of reasons as to why this should not be done, including the privileges of the Deputy GOC. The direction was implemented without affecting my functioning or tarnishing my status. I functioned with just one man, my buddy; no guard and none else. There is huge mismanagement of manpower, simply because they are there for free. Any change that reduces the comfort level or the perceived perks is resisted. We need to grow out of the comfort and egotist zone and be more honest. Every man in the army has to be multitasked. Imagine one man dedicated for the photocopier alone where he is doing nothing at all for ninety five percent of the time, or five runners standing outside five adjacent rooms in a headquarter whose farthest ends are less than fifty metres apart. Later, as a division commander I returned over eighty men and six vehicles attached to my headquarter from units based on the principle of multitasking and efficient use of manpower.

The raising of the division was great fun. We started with borrowed utensils, taking turn with plates to eat. The division was raised in a record nine months and took over the responsibility of eastern Arunachal. We worked hard, played hard and enjoyed hard. In the absence of sports infrastructure, we took recourse to rugby in the rain, my favourite. The entire documentation of raising was handed over to the next division under raising in hard and soft copies, making their task simple.

The administration was streamlined so that the forward troops of the deployed brigade that came under our command, regularly started getting their supplies of fresh vegetables and meat thrice a week, three hundred kilometres from the plains of Assam. They had never seen this before.

Lt Col Krishna, MAHAR, the Lovable Bloke

Except the divisional commander, everyone dined in the mess. Lt Col Krishna, a dark skinny decorated officer, who would be blown by a strong gust of wind, was alcoholic and very chirpy after drinks. A good guy, he was a victim of alcohol. One failed marriage, the second in turbulence. He would often say, I am a superseded guy, what can the Army do to me?

One day as I walked into the mess for lunch, Lt Col Krishna with a drink in hand walked up to me and said,

"One day I will kick you so hard sir, you won't know what has hit you."

I was surprised for sure, since there had been no issue with him. I wondered what led him to behave like this, which for sure was stupid and suicidal.

All officers were stunned and waited for my response. I calmly said,

"See me tomorrow in my office at ten." And walked away.

The legal officer was asked to produce a written warning and be present.

Krishna marched into my office trembling.

I said, "Look, what alcohol does to you and now you are trembling in my office. Where has yesterday's bravado gone? You sometimes say, what can the Army do to you? My options on your state that the Army provides can put you on the road. You know what that means with two marriages in trouble. As an alcoholic, you can be out of the Army and struggle for your pension. The same runner and your buddy will refuse to serve you a glass of water, if they know that you do not enjoy the establishment support. In this remote place, I will give you no work. You will die of boredom, indignity and shame. You are looking to be a time scale Colonel. All that will disappear, if I choose to get after you.

But, I do not look at you as a discipline case. I look at you as a good man who is losing the battle to his circumstances because of liquor. I look at you as a soldier who has earned a gallantry award, who now needs help and get his life back on track.

I am serving you this written warning, but you can count on me to help you, if you come half the way. Now, you have two choices—to continue drinking the way you are and mess up your life and your family's for good, or make a decision to get your life back on track. You make the same mistake again; you will see the wrath of this patient man. You make the right choice, I am with you as a friend."

This man became my Man Friday. He would stand six hours in pouring rain working. When I was leaving the division, he said with a laugh,

"Sir, you had me. No one has got so much work done from me in my whole career. But I loved doing it for you."

Six months after I left, he called me to tell me that he has completely left drinking.

Lesson

Every man is a good a man to begin with. Many become victim of circumstances. Understand them, give them support and dignity and I am sure all of them will respond. Wherever I went, superseded officers were my best troubleshooters. They had no career prospects, but I gave them trust, dignity and responsibility.

The most interesting activity was designing the formation sign. Animated discussions took place every morning with all kinds of suggestions, taking us nowhere. Finally, I directed that no more suggestions. Anyone who has a suggestion must make one and bring one. None came and we finalised a meaningful design in keeping with the role of the division in the land of the rising sun. The Adjutant General refused to approve it because it had three colours. He wanted it made only in two. We were emotionally attached to our design on which we had worked for over six months and we were not giving it up. We waited for him to retire and the division has that beautiful formation sign today.

The Army has the provision for all eventualities; you need the right people who know their job and the courage to see it through.

We had all the fun. We worked hard, enjoyed hard and raised the division as a professional force with the right values, with the motto 'Stand firm – Strike Hard'.

DIVISION COMMAND: INTEGRATION THROUGH FUN

I commanded 22 Infantry Division, called the Charging Ram Division. It was an exhilarating experience. As part of the Strike Corps, we had just one aim—train to be the best.

Our vision, 'To be the best operationally trained division in the country.'

My business as the divisional commander was to build a high 'Human Quotient' of the division, for that would make the difference between defeat and victory. We believed that the success of the Western Command depended on our success, 'the queen of the battle', notwithstanding that the armoured division was discussed more in the sand models, the 'queen of sand model discussions'. We knew, unless we set the stage, by quick initial success, the armoured division will never cross the border.

So, we trained hard. Our philosophy was that we must be prepared to break through within own integral resources. And that required thinking out of the box, empowering junior leaders, and a very high degree of training. It gives me immense satisfaction to state that the division was awarded the Western Command championship banner. And I was sacked. More of that later.

To cite an example of our training standards: All ghatak platoons of the division could cross a seventy-five metre canal in a pitch dark night with full battle loads in less than seven minutes on their own, a majority doing it in less than five minutes. All companies could cross the canal in pitch dark night in less than two minutes in assault boats.

In the Cambrian Patrol competition that we won in the command, a hundred metre water channel was crossed by our team in a flat five minutes. The next best timing in the command was nineteen minutes and twenty seconds. We were proud of the professionalism that was being developed in the division.

As a division commander, my focus was on creating Team 22, an integrated well-knit division and not a division of 22 units. We achieved that by innovative ways such as inter-formation competitions and activities based on the operational order of battle. The brigade teams comprised units that were part of its order of battle and not on its peacetime affiliation. Similarly, whenever the officers and men moved between New Delhi and Meerut, the two stations where the division was split, they always stayed with their affiliated units or formation. Officers and men attended the functions of the units and formations that they were operationally affiliated with. Net result, a well-integrated division where everyone recognised the voices, understood and developed a good rapport with those they would fight a war with.

Our philosophy was 'Think out of the box-keep it simple'. It was derived from my belief that the stereotype will not succeed in war and that surprise was the most deadly weapon of war. The way of life was based on the 'Directive Style of Command' in the true sense. Everyone had the space to practice his art of command within the division commander's philosophy.

I strongly believe in the power of sports in developing the human quotient and energising troops. We organised an inter-formation tradesmen rugby competition and an inter-camp rugby competition. Accepted with reluctance, it became the formation's favourite sport and played every time it rained. Heavier the rain, the better.

There was one football ground in the Signal Regiment which collected almost foot-deep water every time it rained. That was the venue of all fun games. On one such occasion, when a surprise rugby match was organised after a heavy rain during office hours, one officer said to his commanding officer.

"Sir, there is too much water, I think the match will be cancelled."

He was told, "You haven't understood the GOC. You are here because of this water."

Occasionally, I drove to New Delhi to exploit the heavy rains for an impromptu rugby match. The prize was always crates of beer, to be drunk then and there. So, the division could be drinking beer at ten on a working day very often during the monsoons.

I loved these men. My troops in the division camp were from the famous JAT Regiment. When I visited the camp, firstly they were surprised that I was visiting the camp. Then when I asked them what training they were doing, they were stumped. The division camp and training! He must be out of his mind, expecting the camps to do training! They were doing no training. So was the state of affairs in the brigades. I found that shocking. I asked them, did they have a role in war? Yes, they said, they had a role in war. So, if they had a role in war, should they not be training for it?

I directed the division camp that they prepare proper training programme based on their operational role and conduct daily training for two hours every day, at a time of their choosing. The camp commandant, Lt Col Subhash Dhar, was a burly fellow tipping the scales at 160 kilos. A smart cheerful guy, I asked him to make a presentation on the training after a week. As he began, he asked my permission to say something honestly.

I was never shy of hearing anything from subordinates.

He said, "The men are saying, *yeh bara ajeeb sa GOC hai saab. Yeh toh hame bolta hai training karo. Saab ne toh hamari ERE kharab kar di, saab.*" (He is one hell of a GOC. He expects us, the CAMP, to do training. He has ruined our ERE. Posting outside the unit is called Extra Regimental Employment. Expected to be a relaxed tenure.)

The next thing he said was, "*Saab, yeh GOC saab ne ek naya drama suru kar diya saab, Roogby. Yeh Roogby kaun sa tum-tuma hai saab. Saab bolte ball piche pass karo, goal age hai saab, toh goal kaise hoga. GOC saab ko kaun samjhaye.*" (This GOC has started another drama, Roogby. What is this tum-tuma Roogby? He says the ball will be passed backward, but the goal is forward. How will you score a goal then? Who will make him understand?)

The camp got to serious training.

I had my Establishment Officer, another hulk at 120 kilos. Whenever a senior officer visited, he would make it a point to stand next to Lt Col Dhar. In front of him, he looked famished. The visiting officer never even noticed him.

This camp langar cook, also from the JAT Regiment, for whom the Army has given an exotic name, 'community chef', was master at making jalebis, the regimental sweets of the JAT Regiment. He would beat any sweet shop in Meerut. The JATS are fond of churma and jalebi and milk. So, I directed that a three-day jalebi cadre will be conducted in the division. All units will send two cooks to attend. The division found that hilarious. They had heard of all kinds of cadres but never a jalebi cadre. The cadre was conducted, followed bya jalebi competition. The officer in charge had a tough time, eating twenty-six jalebis of twenty-six units to decide the winners. My instructions—on my visit to the units, nothing will be bought from the market and I looked forward to eating their jalebis made in the unit cookhouses. The units delightfully produced top rate jalebis.

The other 'forgotten army' on which I focussed was the Ground Liaison Officers (GLO). The GLO is the officer who with his team is required to coordinate and guide air support. An extremely important asset, he is a forgotten man in most formations and does little of consequence in peacetime. Wonder, how would he perform his tasks unless he trains in peace? I was being briefed by my GLO. The first thing I asked him was:

'What is your task?

He replied. "Sir, I am your force multiplier. Whatever force you give me, I will multiply it."

I told him, "I like your josh. But that alone will not do. I realise that you all need to train and get to grips with your job."

Thereafter, we had a quarterly written and practical test of all the GLOs in the division with eighty percent passing percentages. I suppose we also had the best-trained GLOs in the world!

Our mobilisation practices were conducted most professionally. Every man was retrieved for mobilisation and annual training camp of the battalion from wherever he was, except those who were with the chief.

Every aspect of mobilisation was practiced as the units moved a hundred kilometres away for three days. All teams including ammunition collection parties drove three hundred kilometres to ammunition depots to go over the route, familiarise with the layout of the depot, establish liaison and meet counterparts and ensure the correct documentation for ammunition withdrawal. The unit mobilisation concluded after deployment in the FUP and crossing the Start Line.

The division could mobilise in two hours, notwithstanding that we had two brigades in New Delhi, with all our men except those who were with the chief. My orders were conveyed to everyone who had our man, on some affiliation or directions that these men will be withdrawn without notice in case of mobilisation. Despite resistance from the highest quarters, our men on guard duties in the Infantry Hostel, New Delhi, were withdrawn after a month's notice. They reluctantly switched to security agencies.

While we had this passion for training, our army commander did not share the same passion. When the army commander was briefed by the brigade commander in New Delhi that his battalion would be doing the annual training camp, I was shocked to hear the army commander say, what was the need to do annual training every year. You must send only those who did not attend training last year. And we were the strike corps! Rather than wasting time arguing with him, we did what was right, notwithstanding his directions. His focus was on administration. I pointed out that my division was part of the strike corps and I was required to mobilise in two hours, thus, I must be relieved of mundane duties in New Delhi, which must be shared by 70 other units in the station. I was told by both the army commanders that my brigades had been brought to New Delhi to precisely do that. In other words, I was not to complain about anything and do what was going on. None of them had the heart to question the status quo, even when it was against operational preparedness.

One wonders at the paradox. All talks of cold start et al and your most critical formation of Western Command, meant to set the stage for the strike corps operations, is so heavily committed in New Delhi that does not contribute to the operational readiness one bit. And the Army Commander says what is the need to do annual training camp every year?

My flag staff house became a divisional asset. Every officer and family of the division was my guest in the flag staff house, at least once. There was never a free drink though. Every RAM Warrior was to first climb the rope hanging in the lawn. As the news of the rope trick spread, all commanding officers ensured that their officers passed the rope test. If he failed, he paid for his drink and a call to the commanding officer.

Moral courage has been touted as an important leadership trait. Every senior officer gives a lecture on these traits to his subordinates. Yet, they are the ones who lack this more than others.

"Spineless servility of the staff officers has been the bane of the Indian Military. The case of the Adarsh Housing Society proves it amply. The land in question was under the Army's possession. It was apparent even to layman that the whole project was ill-conceived and murky. Yet the flats were obtained by three Chiefs and many senior commanders by all means, fair and unfair. Instead of cautioning, the staff officers abetted the wrong doings. True to their character, their sole aim was to keep their bosses happy.

Most worrisome has been the cowardice displayed by the Principal Staff Officers (PSO) at the Army Headquarters. The Military Secretary and the Adjutant General are the top-ranking staff officers with immense powers. Yet when a parochial Chief wanted to tweak the process to favour his protégé, neither had the courage to oppose his decadent and wily machinations. Their gutless conduct showed them as unworthy of the high appointments they held.

Similarly, no PSO cautioned the Chief against disbanding the Technical Services Division. Everyone knew it was an act of vendetta."

"Spineless staff officers are a letdown", Maj General Mrinal Suman, *Indian Defence Review*, net edition, 01 Apr 2015.

These are not exceptions. This is the norm.

I have seen many incompetent flag ranks commanders in my service, some of whom have been talked about in the book. My experience as a divisional commander further reinforced my belief in the lack of moral courage and integrity in the majority of senior officers.

One day, the corps commander called me up, very worked up. He said, the army commander is very agitated, saying the Chief, General Bikram Singh, is very unhappy with me, as he has got reports that the commanding officers and the brigade commanders are being called to the division very frequently, leaving them no time to discharge their command functions.

I said, the report, whichever way given to the Chief, is factually wrong. It is incorrect of him to have activated the whole chain without confirming the facts. In my four months of command, these people have been called four times, none of which was avoidable.

I now seek the corps commander's permission to write to the Chief.

He said, "You sure, you want to write to the Chief?"

I said, yes, he needs to know the truth and realise that his sources are feeding him false inputs. The gist of my letter to the Chief is reproduced verbatim:

> *"...I was informed by my Corps Commander, of the Chief's concern, passed down by the Army Commander that the commanders and the commanding officers of 22 Infantry division are being called to the divisional headquarters too often. The information of the Chief is factually inaccurate and the facts were clarified to my corps commander.*
>
> *As a conscientious officer, it is important for me to apprise the Chief of the facts. I also have a bigger reason to write. That deals with the inaccuracy of the report and hence of the reliability of the source dealing with the highest office of the Chief himself. Unverified inputs could become a cause of embarrassment to the Army, if the Chief continues to rely on such sources and hence my letter to the Chief.*
>
> *Since this input relates to my formation, I am certain of the facts. At the time of the Chief's observation, I was in command of the division for four months. The officers came to my headquarter four times, not one of which could be avoided as per military norms. The commanders and the commanding officers were called to the divisional headquarter as follows...*
>
> *To be fair to my commanders in New Delhi, I must mention that the capital, as is well known, has many commitments of great national importance for which my brigades are responsible. They demand*

thorough execution and thus a significant involvement of my commanders, commanding officers and men.

... I give it my hundred percent. My commanders and the commanding officers share my passion and my command is highly motivated and fully prepared to undertake any task with the assurance of success in peace or war.

I would request the Chief to give me an audience to clarify the issue which has been of concern to the Chief."

Sometime later, the MA to the Chief called me to tell me that the Chief had received my letter and he has conveyed that there was no source, it was an anonymous letter. Well, if there was one, which I doubt, is this the way to dispose an anonymous letter? The regulations on anonymous letters are very clear. Biasing the entire chain of command without seeking my clarification on the letter, as is always done on an anonymous letter, is possibly intentional and certainly wrong.

The second incident is a matter of shame for the senior hierarchy.

I had 10 SIKHLI battalion under my command, based at Delhi. There was an incident of mass insubordination in the unit in October 2013, closer to mutiny, a term the Army was shy of using. During this incident, the unit was in the training area, in proximity of Meerut. The troops of the battalion rebelled against their officers creating a very dangerous situation in the unit. This unit was already in a state of distrust between the officers and the men, as it had emerged after my talk with the men. The rebellion was triggered when an enthusiastic young officer, the ghatak platoon commander, manhandled a soldier of his platoon, who had refused to fight in the boxing ring.

The commanding officer, instead of controlling the situation, deserted the unit with his subedar major, displaying the worst form of cowardice and lack of leadership, leaving the unit in a state of disarray and anarchy. Seeing him desert, all the officers ran away too. The leaderless troops resorted to violence, burnt the commanding officer's tents, ransacked the mess and destroyed the officers' and government property. The second-in-command was beaten up and landed in the hospital, as did another officer.

As I learnt of this, I rushed to the unit and found the officers of this unit gathered enroute to the training area. I asked one JCO to get the men together for me to speak to them. I addressed the men, many in an inebriated state.

I heard them for fifteen minutes and then spoke to them. I appealed to them that they had written the name of their unit in black that will be remembered forever along with other units where such incidents had happened. I said, I am here to save the unit's honour, not for my sake, but for their sake. That this matter, while very serious, will not leave the boundary of the training area. I asked them when they get back to the units, when their wives will ask them, "What have you done to your unit's name", what will you tell them?

They realised they had made a great mistake. They agreed that this matter must be contained in-house.

As I was saying this, I got a call from home to say that the news of the incident of this unit is being flashed all over on many TV channels.

I told the men of this and said there is little that I can do to contain the damage in-house anymore. Clearly, some soldiers had leaked the incident to outsiders, and in no time it was public information.

During my continued efforts to restore order and the authority of the commanding officer, the men in unison refused to have this commanding officer back. Very bluntly, they said, he is not acceptable to us and we will not have him back. They said, of this lot, only Captain Chikara can come back to the unit. The men went back to their tents and had their food. When I told the commanding officer what his men thought of him, he looked surprised, stating he was born and brought up in this battalion. Many like him have a false sense of their standing with the men.

So much for the confidence the commanding officer and the officers enjoyed amongst the men.

Rugbying the Battalion Back to Order.

The next day, it was pouring. I had the battalion fall in and got all the officers of the unit. After I and the brigade commander spoke to the men, the commanding officer apologised for the conduct of the young officer. The next one hour was a no hold barred rugby game, with me as the GOC playing in ankle-deep slush with the combined team of officers, JCOs and the men of the battalion. General Campose, the Army Commander, spoke to me on phone, asking me to be cautious. I told him, I am in the middle of a rugby game with the men and what happened last night, is history.

The rugby was followed by some fun games in pouring rain and the battalion was back on its feet and training the next day with the unit command authority restored.

Based on my personal witnessing of the events in the unit and the shocking display of cowardice and the command failure of the commanding officer, I recommended that he be removed from command and also not be given second command of a battalion. My recommendations were also endorsed by the corps commander.

Interestingly, subsequently, the Deputy MS to the army commander, then Brig Rambir Mann, called me up to convey the army commanders' strong reservations against my recommendations on the CO 10 SIKHLI and wanted me to revise my recommendations.

"On what grounds, may I ask?" I enquired.

He said, the army commander says that the officer has a good record of service, and my recommendations will ruin his career.

I said, "The officer has failed the basic test of leadership. He has displayed not only absolute failure of command, but also cowardice of the most shameful kind, running away from the men he served with for decade-and-a-half. What record of service are we talking about?" I told him, "I am not convinced and cannot consider this at all. What is the second point?" I enquired.

He said, "The army commander says that you have convicted the officer without the court of inquiry being completed, hence your recommendations are not justified."

I said, "I have made recommendations as required vide Defence Services Regulations Para 120, and they are independent of the court of inquiry. They are based on my personal witness of the events and not on staff officers' recommendations. The court of inquiry has its own mandate and is in progress. If the court finds anything contrary to what I have stated and recommended, the army commander, as the superior commander has the authority to overrule me, if he has the justification and I will have no problem with that."

I told him, I was fully convinced by my recommendations and cannot morally bring myself to change them. He can convey my view to the army commander.

After I relinquished the command of the division, the corps commander was pressurised to change his earlier endorsement of my recommendations. Sadly, with nothing at stake, he buckled under pressure. Eventually, no one in the battalion was given any serious punishment. The commanding officer was given second command.

General Bikram had spoken to me after the incident and conveyed that strict action be taken, but in the end, it was business as usual for the commanding officer despite grave command incompetence.

In context of the action taken against the commanding officer of an artillery regiment, who had a similar incident in the unit at Leh, the disposal of CO 10 SIKHLI was a poor reflection on the judgement of our senior leadership and set a wrong precedence as regimental affiliation was allowed to overshadow their rational judgement.

My issues are not so much with what happened in this unit. My concern is with the way the command hierarchy—the commanding officer of the unit, the corps commander, the army commander and the chief—responded on this. The commanding officer ran away. The corps commander changed his initial recommendations, apparently under pressure. If the army commander genuinely believed what I was conveyed to by his Deputy MS, I would expect him to overrule the subordinate commanders, if he believed in the justification he was propagating, rather than ask his subordinate commanders to change their recommendations and fire the gun from their shoulders.

If he was doing that at the behest of the Chief, then I would have expected him to have the moral courage to tell the chief that this was too serious a case to be let off on regimental association.

In the whole chain, I was the only one who had a lot at stake, and I was the only one to stand by my conviction and suffer in the end, for which I have no regrets, whatsoever.

My division was awarded nearly ninety awards in peace and won the Western Command Championship Banner for all round excellence for the duration of my command and I, the divisional commander, was found unfit for promotion. That is of little significance in my scheme of things.

Why are the MES assets so poor in quality when we have major stakes and control the projects ourselves?

After I took over the division, the construction of the new divisional headquarter building (KLP) commenced. Within a week, I stopped the construction because of non-compliance of the terms of contract. The contractor, Mr Dhall, pleaded that this has never happened in his fifty years of construction business and it will bring him disrepute.

I said, he should have thought of that before and not a brick will be laid unless he complies with the terms of contract in toto. The next day the commander works engineer (CWE) walked up to me to tell me that I did not have the authority to stop the work.

"So, who has that authority?" I asked.

"The Chief Zonal Engineer," He replied.

"Where is he based?" I asked.

"At Bareilly," He replied.

"Who knows what is happening on ground and whose reputation is at stake? And who is the user?" I asked.

The CWE stood speechless.

I then told him, "It is my headquarter being constructed and my reputation is on the block. Even the chief cannot tell me to resume the construction unless I am satisfied. Get that straight and go and tell anyone whom you wish to tell what I have said. Not one brick will move unless the terms of the contract are met."

The next day I called the contractor, the CWE and all the associated MES staff and apprised them of my position. I told the contractor, fulfil the terms and you can start when you are ready, or else wait for my tenure to finish. I said, not one penny out of this fund will be diverted anywhere. If anyone says that commission has to be paid to the divisional commander or anyone else, walk up to me immediately, whatever time of the day or night. Any violations, a court of inquiry will start in less than twenty-four hours.

I also told the contractor and the Garrison Engineer (GE), that if the building is completed in my time, it will be inaugurated in full media glare by the contractor, Mr Dhall and the GE, then a Major. There will be no chief, army commander or the corps commander called to inaugurate this KLP. The inauguration plaque will carry their names.

The next day, Mr Dhall walked up to me in my office. He said that he will start construction in seven days after fulfilling all the terms.

Then he told me something I was not expecting.

He said, "After hearing you for the last two days. I want to make a special contribution to this project. My profit in this contract is seventy lakhs. I will contribute twenty lakhs from my profit to this building."

I said, "Why do you want to do it?"

He replied, "I see sincerity in your purpose and you are the second officer of this kind I am seeing in my fifty years of business. When you said, you will ask me and the GE, a Major, to inaugurate this building and put our names on the plaque, I decided that this will be the best building that I have made."

He actually spent more than twenty lakhs of his profit on the building including a massive central dome that lit up the building with sunlight, covering two open corridors of sixty feet by forty feet by concrete roofing and two big rooms adjacent to the operations room and sand model rooms, not part of the approved and budgeted structure.

Unfortunately, the building was not completed in my time and it was inaugurated by the army commander.

Lessons

The best things are done when the purpose goes beyond your own self. MES and the contractors will deliver if you mean business.

My last command assignment, the 22 Infantry Division, in 2013, was the icing on the cake of this wonderful journey of soldiering.

As for me, no commander could be blessed with better concentration of units, officers and men in one division as I was. They surpassed my enthusiasm and beat my passion. They raised my stature with their dedication and performance. I had the full confidence in their courage, commitment and the skill to deliver in battle. Leaving the division was as heartbreaking as was leaving my battalion.

I broke the pen with which I wrote my farewell remarks on the VIP Book. The broken pen stays as a memory of the glorious association with the CHARGING RAM DIVISION.

Command in the army is all about leadership and moral courage is the mother of all leadership qualities. It makes the rest possible. Courage in leadership means the following:

- It is about guts to stand behind decisions when you know that you are the one who will take the fall.
- Courage is not the absence of fear. Courage is the ability to move forward in the face of fear.
- Courage requires having a voice—a voice that addresses difficult subjects and delivers tough feedback to the superiors and shares dissenting opinions from the subordinates.
- Courage is doing the right things even when it is dangerous to do so.

- Courage requires letting go and letting people do their jobs
- Courageous leaders do not make excuses when they err—they apologise.
- Courage takes risks that others are not willing to attempt.
- Courage addresses the conflict head on.
- Courage challenges the status quo and makes changes before change is forced upon them.
- Courage protects his flock when they have followed your orders.
- Courage takes the blame and gives the credit.

* * * * *

Defence Attaché: All That Glitters is not Gold

I served a tenure in Turkey as the Defence Attaché (DA), accredited to Syria and Lebanon from 2002 to 2005.

Turkey is a beautiful country with amazing historical places, truly a tourist's delight. Ninety-eight percent Sunni Muslims, it is modern with a Western orientation and a secular outlook, striving to join the European Union. The Turks are a proud people and nationalists, and take immense pride in their language, culture and flag. The western part is more West-oriented and the eastern part more conservative. The Turkish military is very European in its orientation. During the Ramazan, there were no dilution of the routine in the military, while everyone was free to fast. Prayers were permitted only inside four walls of a room or in a mosque.

Strict discipline is enforced. Every Turk believes he is a born soldier. With conscription, everyone serves some time in the army. Military service is considered a national duty, during which they were not paid their salary, only a small subsistence allowance. The effect of this is seen in daily life as people are generally well-disciplined and follow the civil code of conduct. The secular nature of Turkey is being seen to be marginally affected after the current AKP Party, considered an Islamist party, won three consecutive elections with a majority and upturned certain practices. The military has lost its prominence over the last decade under this government.

Attaturk remains a demigod, engineered into a larger than life icon for the Turkish military. Many practices were in vogue to keep his legacy alive. A sprawling Attaturk Mausoleum used to be the most sacred place in Turkey. Once a year all the cadets in the military academy and soldiers in Ankara would run to the Attaturk Mausoleum to remember him and keep his legacy alive. The President reviews the passing out parades of the three service academies every year without fail. It is a country which recognises the role and importance of its military. Compare that with India. I do not remember when the President last reviewed the passing out parade anywhere.

All was not perfect with the practices of the time. There were resentments on application of secular principles, such as not allowing girl students wearing headscarves in government colleges and universities and government jobs. The secularists were so staunch that during the national Independence Day dinner at the Presidential Palace in Ankara, the defence minister and the prime minister of the ruling AKP Party were invited alone because their wives wore head scarves, considered a symbol of political Islam in Turkey.

Barring the problems arising out of the Kurdish insurgency, Turkey was a very safe place then. Women wore Western-style clothes. Girls returned late from work and the society was permissive. The number of girls smoking seem to outnumber the men. Girls and boys mingled without anyone raising eyebrows. Liquor could be bought anywhere. Yet, there was discipline on the streets and public places.

Turkey has kept out of all wars after the First World War. It kept out of the Iraq War 2003 too, despite being a NATO member and an ally of the US and bordering Iraq. The Turkish parliament did not allow the US troops to transit through Turkey to northern Iraq, even though the US infantry division had disembarked on the southern port of Incirlik and was heading for operations from northern Iraq. They allowed the use of their bases and ports to the US military for logistics. However, it drew Turkey into terrorist sights as they targeted British and US interests in the country. A member of NATO, Turkey remains a bridge between the East and the West, yet there are major issues relating to Turkey's ambition to join the EU. For the present, it looks a distant dream. In my opinion, there is one reason that justifies Turkey's EU membership. In the world engulfed by Islamic extremism, Turkey, a moderate Muslim country, can be an example to show that West and Islam can coexist peacefully. Lately, Turkey has been hit hard by the developments in the Middle East and the immigrant crisis in Europe. Geography has destined Turkey to play an important role in this century. Its internal dynamics will only be aggravated by a referendum for the Presidential form of government with sweeping powers to the President held in Apr 2017, seen by the sizeable population as taking Turkey towards an authoritarian system.

The Turkish military were great hosts. They enjoyed a position of prominence in the government. The incoming DAs were received at the aircraft gate by the protocol officer and taken through the VIP lounge without a fuss. If a DA informed the dealing office that he would be moving out of Ankara, he would be guided through the VIP lounge without going through security check. Any visiting military dignitary was extended admirable hospitality.

The Indian Director General of Military Intelligence, who spent a day's lay over at Istanbul, was surprised by the hospitality extended to him, even though he was not on an official visit to Turkey. He commented, "We can never reciprocate this hospitality".

The Turkish Chief of General Staff (CGS), a five-star General, stands fourth in protocol, ahead of the defence and foreign minister, and that cannot be disregarded simply because we have a different system. When General Hilmi Ozkok, CGS, visited India on a reciprocal visit in 2006 after the first-ever visit of the Chairman Chiefs of Staff Committee (COSC) and the Air Chief in 2004, I learnt we cut a sorry figure in reciprocating with basic hospitality. His entry to the Taj Mahal was not without embarrassment and his aircraft was kept on the runway in Mumbai for an hour and not allowed to deplane on some trivial issue. What a shame. It can only happen in India; the 'chalta hai' attitude and lack of commitment prevails everywhere. The responsibility has to be taken by the MOD, the MEA and the Military Intelligence. They have to understand that such visits are not army visits; these are bilateral visits involving two countries. When we fail, we do great damage to our own reputation as a nation.

I had an excellent equation with General Hilmi Ozkok. He was an excellent person and a highly respected General in Turkey. Sad that we let ourselves down and he had to convey his displeasure on return from the visit.

Turkey and Pakistan have excellent relations on all fronts. Given this fact, the first-ever visit by a Turkish CGS did no good to our cause. The problem is that military visits are seen by other ministries as the services' baby, with little or no concern from others. There is no single point coordination agency. Thus, it is left to the service to tie up with a plethora of ministries that are involved; MOD, MEA, Ministry of Aviation, Airport Authority of India, Ministry of Tourism, etc., who are at best indifferent to military visits.

I remember that when the Indian Defence Minister was scheduled to visit Turkey in 2005, the Ambassador was indifferent entirely, as opposed to a visit by the speaker of parliament a few months ago. With just a few days to go, there was no word of the visit either way. The Turkish Ministry of Defence Liaison office repeatedly contacted me stating they needed to plan a lot of things including the availability of their service chiefs.

The ambassador, despite my repeated efforts did not budge and refused to contact the MOD on the visit, stating, "Why should I call them, they will tell me if it is cancelled." This reply from the ambassador, three days before the visit, apart from being frustrating, spoke volumes of the kind of people who head our missions and represent our country. Finally, I told her that I am telling the Turkish Liaison Office that there is no visit. At this, she got alarmed, saying, "How can you do that?" I said that you refuse to make a call, and I cannot let the country look foolish in Turkey. I will face the consequences. I informed the Turkish MOD that the visit is off. Fortunately, the visit did not take place. Similar was the story during the visit of the Chairman COSC, Admiral Madhvendra. The dealing Admiral in Naval Headquarter was running from pillar to post to get the government approval. I then sent a signal and rang up the Indian MOD, telling them that if I did not get an answer by 1 pm the next day, take the visit as off. Pat came the approval well in time.

This has been and continues to remain the practice. The Army must demand to be treated with respect. You cannot treat service chiefs in this manner. The Service Headquarter must insist on an annual approval of the chief's visit and there should be no cancellation except on extreme political or diplomatic grounds, well in time. A last minute suspense on the visit causes extreme stress on the DAs and shows our country in poor light abroad.

When the Chairman COSC, the Naval Chief at that time, visited Turkey, his wife mentioned that they were a little anxious and sceptical, wondering how much interest the DA will take, being from the army. That is a sad commentary on the services. Possibly, the Naval Chief had an experience where officers from other services did not respond with the seriousness that they should have, simply because he was not writing their report.

At any time, there were over sixty Pakistani military personnel in Turkey doing some courses, etc. There was a pilot exchange programme vide which they exchanged fighter pilots in their training academies. A large number of Pakistanis were studying in Turkish universities. Musharraf had spent some years in Turkey as a child with his father posted in the Pakistan embassy. He tried to emulate Attaturk without success.

I invited a Turkish Colonel, a former DA to Pakistan, to my place. He said, when a Pakistan DA invites a Turkish officer, there are no questions asked. But when an Indian invites one, he has to explain how he is known to him.

Pakistan was promoting cricket in Turkey and would organise matches. I was invited by the Pakistani DA to play. I reached the university venue at the given time. There was no sign of the Pakistan DA for the next one hour. His JCO PA met me and I said, I have been waiting for a long time, there is no news of his Colonel.

He said, "Sir, Pakistani officers are decorative pieces. Can't expect much from them. I have come from the same location as him after changing four buses. He never offered to bring me in his car which is a government transport. Sir, I can tell you if the Pakistani soldiers and Indian army officers were part of the same army, no one could touch them." This remark was also made by a Pakistan JCO in the 1971 war in Bangladesh, when he saw that the Indian officers also spent the night out in the cold without blankets like the troops.

When PM Vajpayee visited Turkey, the reconnaissance team landed there. PM Vajpayee had problems walking. It was a funny sight to see this team measuring with a tape every inch of the distance that he would walk. Surely, there are smarter ways to pace up the distance. All official visits in Turkey start with a visit to the mausoleum. There was no way the PM could negotiate the large number of stairs and manage the distance, given his problems. The MEA's efforts to get the Turkish military to permit the use of a golf cart in the Atatürk mausoleum failed. The MEA then approached their ambassador in New Delhi without success. Eventually, the ambassador asked me if something could be done. In six hours we had a positive decision.

The Turkish General told me to go the mausoleum at a given time. A Turkish colonel was waiting with a frame, the size of a golf cart. He said, here it is, use the golf cart wherever it can go. The ambassador was surprised that I had the clearance.

Later in the hotel, a MEA Joint Secretary was loudly boasting how he got Turkey to agree on the use of the golf cart by just one phone call, when everybody else had failed in the embassy in Ankara. Much against my nature, I put him in his place right there. PM Vajpayee walked the fifty metres of open space and then was taken in a wheel chair, away from the media glare. So much for the efficiency of the ambassador and the MEA.

The DA is considered a glamorous appointment, but it can be a tough one too. Former Vice Chief of the Army Staff, Lt General Pattabhiraman, who was also a DA in Turkey, described the posting as a good station but a tough assignment. I could not agree more. Having seen three embassies for over three years, one feels saddened at the way some of these embassies function and are a drain on the exchequer achieving little worthwhile.

As military men, you live by excellent military values, where the organisation comes first. Such values were conspicuous by their absence in the embassies that I served. I was appalled at the lack of concern to the organisational goals at all levels. It was my impression that foreign service officers are poor team men, self-oriented, class conscious and live for their personal growth. Ambassadors either were incapable of providing the leadership to forge a team or didn't care a damn about such things. It is my impression that a large percentage of foreign service officers lack the grooming for the important job their service is expected to perform.

Ego was the hall mark of the foreign service officers and their wives. As service officers, one is posted there with over twenty-three years of service as Colonels or Brigadiers. Some moron has equated the colonels with first secretaries of the MEA, who become first secretary at six years' service. This must be set right.

During my tenure in Turkey, I served with two ambassadors, one a Bengali male and a lady ambassador. Turkey is a small mission. Five officers in all, including the ambassador. One out of the four officers was also a Bengali. The parochialism of this Bengali ambassador was so strong that in formal weekly conferences, the ambassador would speak to the rest of us in English and converse with the Bengali officer in Bengali. One found it astonishing that such important positions were entrusted to such small-minded people. I knew that I would find it difficult to get along with him.

In one instance, this ambassador refused to stay in one hotel, because it was below his authorisation, even though the ambassadors of Canada and South Africa were comfortable staying there.

One admires the Indian government which looks after its officers in foreign missions so well. I guess the bureaucrats look after themselves well, for they are the ones who make the rules. Almost everything is paid for, expensive children's education, good accommodation, car insurance, car transportation, servant wages, unbridled medical and travel expenses, hospitality allowance, etc. The ambassadors have it laid out for them. There is very little that they pay for, for example they paid just five percent of their electricity and water bills, while the other officers pay full. There is no limit on the house rent. All told, the embassies run the country dry of foreign exchange.

The ambassador in Lebanon at the time, apparently rarely contacted the Lebanese foreign ministry. At the end of his tenure, when he wanted an appointment to be fixed with the foreign minister for a farewell meeting, the Lebanese officer commented, "We did not know you had an ambassador here."

When he saw me making serious efforts to improve the bilateral military relations, he advised me not to waste my time and energy.

A DA's life can be difficult. I was told by a MEA officer that there is an unwritten understanding in the MEA to keep the service officers under their thumb and give them no liberty. They are seen as outsiders, who are unwelcome on the MEA turf. They are kept out of the loop, except when some difficult work is to be done.

When I rang up the deputy head of mission in Syria to seek his advice on my first visit, he retorted, "Colonel, do you realize that I am not your counterpart here?"

I said, "Well, this is my first visit, I thought I will speak to you and seek your advice. But if this is your approach, and you think I have made a mistake calling you, then it is my last conversation with you." I was asked by this counsellor to address him as sir. I told him to hop, as the right mode of addressing any civilian officer by an army officer was by his appointment or as Mr X, etc.

When I visited Syria the first time, I refused to meet him, to set the rules of engagement right. He then complained to his ambassador, who wrote a letter of complaint to my ambassador in Ankara. I told him what had happened. He also quipped, "Ego is the biggest problem in MEA."

The ambassador in Syria was sozzled by eleven in the morning. The one in Lebanon made no efforts to accomplish anything. The chancery was in shambles and the staff disinterested.

On one occasion when I was being advised to do certain things by the ambassador in Turkey, I said, "I am an army officer and not a MEA officer."

The ambassador said sarcastically, this is not your army colonel. I replied, "I am an army officer and my code of conduct does not change, no matter where I am or what I am doing."

Over a period of time, differences emerged between me and the ambassador. When he was to initiate my confidential report, over six months overdue, he sent his PA to get my signatures on the blank annual confidential report (CR) form. I knew what that meant; he would have nothing good to write about me.

I told him, as a PA he must understand what it means and advise the ambassador. He said, he was doing what he was asked to do, and the ambassador has said he does not want me to know what he writes. I told him to tell the ambassador to hop and tell him that the DA refuses to sign the blank report form, as it is against all norms which he very well understood, having written reports of the DA earlier. Later, I told the ambassador that he was violating the government instructions on initiation of the CR. He was free to write what he wishes, but it is unimaginable for him to ask me to sign on a blank form, and the CR itself was invalid without my signatures. When you deal with such people, confrontation can't be far behind. My CR was initiated thrice during my tenure, and thrice it was never shown to me. It is clear what they would have written. Why so, one might ask? Was I a trouble creator? It was for one simple reason that I was not prepared to compromise the position of the Indian Military and I have no tail to wag.

Eventually, we had a showdown. He organised an 'Indian Bazar' in the embassy premises. It led to a major confrontation between the wives of the ambassador and the deputy head of mission on money matters. The second lady, unhappy that she was not paid the appropriate amount for the artefacts that she had put up from her house, threw the money on the ambassador's wife's face and walked off. She circulated a letter accusing the ambassador of violating all regulations and making money out of this Indian Bazar. The ambassador, petrified and out of his wits, called all the officers and families to the chancery and spoke to them, asking them if they thought he had made money from this event. One of the ladies, ambassador's confidante, went on a signature spree to prove that holding the bazar was a consensus decision of all the ladies. When she came to my office to canvass for my wife to sign the letter, I told her to leave my office and stop ganging against one person. I also conveyed my displeasure on the value system in the embassy.

The next day, the ambassador, quite agitated, asked me to meet him in his office to discuss some serious issues. I guessed what that was for. I was right. When I went in, I was surprised to see the ambassador's wife too in the office, both in a pensive mood.

He said, "I have very serious issues to discuss, and I would like answers in yes and no."

I said, "I know what you want to discuss, so my answers will not necessarily be in yes or no and they would need some explaining."

"Did you tell Mrs X that I have stolen money from the Indian Bazar? Answer in yes or no," he said.

"I have no justification to say either yes or no. I must tell you that, sadly, I learnt of the Indian Bazar from the ambassador of Pakistan and the

Bangladesh DA and not from our own ambassador. The Pakistani ambassador asked me where was this money going to. I said, I have no idea, as I don't even know that the Indian Bazar is being organised. I certainly told Mrs X, that I had no idea where this money was going till the final day when I visited the Indian Bazar."

The ambassador attempted some lame excuses that he had discussed the idea with the Pakistani ambassador to seek his views. I told him, that he would rather discuss any idea with his own officers rather than the Pakistani ambassador of all people.

Our discussion carried on for long during which the ambassador's wife butted in angrily. The ambassador then asked me if I have passed aspersions on his leadership. To this, I replied that yes I did. It was my belief that in the three embassies that I am accredited, no one has his heart in ensuring that the organisational goals are achieved. Everyone is an individual pursuing his own personal growth. And when I compare this with the Chetwode motto, which is the guiding theme for the Indian Army, "The country comes first always and every time, your men come next and you as officers come last...." The embassies that I served with were a shame.

Finally, he again asked me if I thought the ambassador has stolen money from India Bazar,

> I replied, "I have no basis to say either yes or no. I do not know whether you have followed the regulations for the conduct of such an event. The question is, why is there so much resentment after the event? But you as the ambassador have made a major mistake, which, in the military environment, would have meant serious trouble."

"What mistake have I made?" he asked.

"As you say, the money was raised for the old age home in Kolkata. The problem is, there is a definite personal interest which neutralises all the goodwill you may have had. And that is, Kolkata is your home town. You belong to Kolkata."

The ambassador was taken aback and gave lame excuses as why he did it for Kolkata. I said, it is important how others see it and not how he sees it. I told him my impression of what I thought of it. Finally, I had to tell him, that to me he always looked more like the ambassador of Bengal rather than the ambassador of India.

The Army itself is of very little assistance to the problems of the DA. Before I left for Turkey, the Vice Chief, Lt General Vij, told me, "that while we do not easily jell with civilians, I do not want any report coming against you. I already have two reports against the DAs from two ambassadors."

It was quite amazing that those days, officers were sent as DAs with no orientation training worth the name. One learnt whatever one could on his own. Many countries run a three months' orientation course where all issues related to their functioning are covered, including issues such as how to deal with a hostage situation, etc. I made comprehensive recommendations based on my experience to make the DAs more effective, only to find that my letter was not even traceable in the concerned office in the MI Directorate. Despite the HQ Integrated Defence Staff (IDS), there is no single point of contact for the DAs. Respective services are continuing to guard their turf which makes the functioning of the DAs more complex and inefficient.

My stint with the second ambassador started on a good note and stayed that way for some time. I observed that the MEA officers posted there lacked the basic courtesies that must be extended to a female head of mission. She once asked me, what I thought of the MEA officers? I said, by the military standards, most would do not qualify to become officers. She asked me why.

I said that being a good officer is not about the ability to cram reams of paper and reproducing them in the exam. It is about character and personal qualities. The officers that I have dealt with in three embassies lead me to this general assessment; they lack major traits of a good officer; leadership to run a motivated team, commitment to the organisational goal, selfless service and teammanship.

Surprised by my frank assessment, she smiled and said that she was their instructor in the Civil Services Academy, Mussoorie. I told her that my response would not have been any different, even if she had told me that before.

Gradually she started eyeing the defence wing funds and demanding that I share the expenditure of my discretionary grant with her. All expenditures of the defence wing are controlled by the embassy. The only fund the DA has is the discretionary grant, at that time about USD 200, for a month. I told her that I follow instructions to the core where it comes to this grant and report to the army headquarters and unless the instructions are amended, I cannot share the details with her. She then told me that her DA in Sweden, in her last mission as the ambassador, was doing it. I told her that he was violating instructions to earn your favour and setting a wrong precedence.

Precisely, because of such self-serving officers, the system suffers and others are placed in difficult positions, as I am today.

Eventually, when I declined to make the defence wing funds available for expenses for which they were not meant, she passed a written order, directing that thirty percent of the central heating and electricity charges will be paid by the defence wing. I replied in writing that the ambassador has refused to discuss the matter with me four times and the order is in violation of the laid-down policy and not justified. Given the fact that the defence wing has only two persons including my army PA, we account for negligible use of heating and electricity expenses of the embassy; I regret my inability to comply on merits of the case.

The ambassador blew up when she saw my noting and said that no one has ever written such a noting in her career. She said that I was creating a tamasha, and that I would not do so if there was a man ambassador. She threatened to have the defence wing closed.

I told her, she had forced me to reply in writing in response to the office order issued in writing, having declined my four attempts to discuss this issue. As for being a lady, she could check with her male predecessor as to what I had told him when he created an issue, sitting in the same office. As the ambassador, she had the same powers as a man, and she could call the Army Chief right then to report on my actions and initiate her recommendations to get the defence wing closed; I was prepared to leave the next day.

I told her that in my service, I have seen many soldiers die unnecessarily because some commander did not have the courage to stand up to his superior. Since then I have decided, I will always speak my mind, no matter the consequences. No one is dying here, but my views cannot be disregarded. I cannot be treated as you do your first secretary. I represent the Indian Military and you will have to deal with me accordingly. Then I told her, I respect her views and will send her recommendations to the army headquarters for their consideration, with my recommendations that they are entirely unjustified for all the reasons I have given in my noting.

She was taken aback. I said I can't be disloyal to my organisation. The matter was dropped for the balance of my tenure.

Sometimes, ambassadors work at their own whims and fancy. I requested for a vehicle to pick up my wife from the airport about thirty kilometres away. The vehicle was denied on the grounds that my wife was not travelling on a government-paid ticket.

I asked the head of chancery, was it the ambassadors' direction, to which he said yes.

I walked up to the ambassador and told her that I have sold my car pending posting and my move has been delayed because of my successor's delay. This was the first time I have asked for a vehicle and I believe you have denied it.

She said, "Yes, there is too much of misuse of vehicle by officers."

I said, "Are you including me in this misuse? Please quote one instance when I have asked for this vehicle for personal use."

She said, "You use the vehicle the most."

"My nature of official commitments and duties demand more use and the vehicle is demanded for a legitimate reason, and through the laid-down procedure. How does this become misuse?"

"I have decided that the vehicle will only be given strictly as per rules and for official purpose only." She said.

I said, "That is a good policy, but you will not be able to implement it."

She asked "Why not?"

I replied, "You sent the audit team for three days' tourism at embassy expense in hired cars. Your own guests will come too. You have a daughter too. How often has she used the ambassador's car, and how are all these more official than a vehicle to pick up an officer's wife from the airport?"

She did not know what to say.

When my successor arrived after about three months, I offered to bring him and introduce him to the ambassador. She declined, choosing to meet the DA designate alone. After his meeting, he said, the ambassador wants the visit of the Director General Military Training to be arranged. I told him, the ambassador has to speak to me, I am still the DA. I told him, this visit would be a waste of national money and I would not let that happen. It did not.

After the handing–taking over period was over, I reported to her on phone and said that I was leaving the next day. However, if she wishes, I could bring the new DA to her office and both of us could give her the handing–taking over report or she can meet the new DA alone, as she had done a week earlier.

She asked me to come with the DA designate. She complimented me for the work I had done. She then asked me how was my experience here as the DA?

I said, "Madame, you do not have to ask me for formality sake. You know, I speak my mind."

She said that was okay.

I replied, "My experience is based on dealing with three embassies for over three years. It can be described as good, bad and ugly. The good part is because of the warmth, hospitality and experience with the Turkish Military and the eighty-five defence attaches from sixty countries. We were like one family bonded by common military values which made my stay wonderful.

The bad and the ugly part was because of the Indians in the three Indian embassies from top to bottom. In this foreign land, the biggest strangers to me were the Indians from the MEA with big egos. Given by the work being done, it is my conclusion that some of these embassies can be wound up without detriment to the national interest."

She was taken aback by my reply and attempted lame justification.

My tenure left me thinking, how much the military sacrifices on values of selfless service, loyalty and dedication, values which I found conspicuous by their absence. I wondered, as the commanding officer, how worthwhile was it, combining activities and transport to save the country some extra litres of fuel, putting troops to inconvenience, when foreign exchange just flows in the embassies with little output.

Air India, the unproductive Maharaja. PM Vajpayee visited Turkey during my tenure. I was surprised to see the numbers of free loaders on the trip. In particular, hordes of media and MEA personnel. He spent two days in Ankara and two in Istanbul. There were two groups of maintenance parties for the PM's aircraft, each between sixty to seventy persons. One group each for Ankara and Istanbul, staying there from three to eight days in five star hotels. On one occasion, I accompanied a friend who was meeting his Air India pilot friend in Dubai. Well, for a half-day halt, he was in a seven star hotel. Compare this with private airlines who do the same job with much lesser; not to compare with our fighter pilots and helicopter pilots who have no such luxury but perform the most dangerous flying.

I wonder who is concerned for the taxpayer's money? Everyone is making hay while the sun shines.

The tenure was a great learning on all counts. A DA deals with the highest levels in the host country and can make a difference in the perceptions of the country.

Once during a discussion with the Turkish defence secretary, I mentioned that "Turkey has brotherly relations with Pakistan. Despite suffering from terrorism yourself, you turn a blind eye to Pakistan's involvement in terrorism in the world."

He said, "Yes, the Turks are emotionally grateful to Pakistan. They contributed to the Turkish war of independence. The Pakistani women sold their jewellery to support the Turkish war of independence."

I said, "That was from 1921 to 1923, and where was Pakistan then? Pakistan was created only in 1947 after India's partition. So, you need to understand that assistance came from India and not Pakistan."

That really hit him. He said, "Gosh, I never realised that, neither do the Turks. Everyone here believes that. This is the first time I am hearing this and I realise it. But what has your embassy been doing here?"

So much for our embassy in Ankara which has been there since the early fifties. It takes a DA to do what well-paid diplomats should be doing. I told the ambassador about my discussion and the need to put the record straight. She too was surprised.

There are funny incidents almost every day in life and the embassy was no exception. The functioning of the embassy in some ways was as much tragic as it was comical.

As an army officer brought up on the best values of work culture, camaraderie and selfless service, one often feels out of place there.

A lady Second Secretary (PA) had one kidney. She had served in the US, Turkey and Australia in succession, in all, over eleven years at a stretch. Her monthly medical bill was a fortune, but who cares. When it comes to work, she grabs an office on the second (top) floor, next to the office of the Deputy Head of Mission, also located there. Her boss, the Head of Chancery, is on the first floor. The secretary of the Deputy Head of the Mission, on the other hand, is on the ground floor. So, as you can see, no one is where they should be. The Ambassador simply does not find it worth addressing the problem for reasons best known to him. She refuses to come down to take instructions or dictation citing her single kidney. But she is happy to climb up to the second floor because she wants to be with the official ISD phone, away from others. So her boss, the Head of Chancery, either types the letter himself or walks up to his secretary on the second floor to deliver and collect the draft. When the Deputy Head of the Mission on the top floor has to get a draft typed, he phones his secretary on the ground floor. The secretary, not keen to take the long walk up the stairs, declines and waits on the ground floor lobby. The Deputy Head of Mission tosses the paper that flies down like a 'katti patang'. Entertainment 'gratis', it was like an organised street play as the locals watched with amusement.

Little knowledge is dangerous, as they say. Even more dangerous are half-baked language skills. Before I went to Turkey, I was given a very basic exposure to the Turkish language. Eight periods in eight weeks by a retired professor of Turkish history. Defining my language skills as less than rudimentary would be a big exaggeration. The Turks are very nationalist and proud people. They take pride in speaking Turkish even if they know English. I spoke to the Turks in Turkish for better integration, never mind my skill level. They realised that and would accept my mistakes. There also was a DA from Turkish Cyprus. We were good friends and got along well. He spoke decent English and did not insist on speaking Turkish. I met him during a reception after a long time. I chose to speak to him in Turkish.

"Cok zaman, size bakmadim, Ozkan Bay," (Long time did not see you, Ozkan), I said.

"Evet, Ben tatille gitti." (Yes, I was on vacation).

"Cok Guzel,"(Damn good), I said.

"________________" replied Ozkan in Turkish.

"Masha-allah (Great)," I said.

"________________" replied Ozkan in Turkish again.

"Cok Guzel," (How wonderful!) I said again.

I failed to notice the changing expression on his face. Ozkan must have been wondering something is wrong with me.

He said in English, "You are not understanding what I am saying."

I said, "Really!"

He then translated the conversation between us.

"You said that you had not seen me for a while. I said, I had gone on vacation. And you said, damn good. Then I said, I had gone because my sister had died and you said, excellent. Then I said, we buried her yesterday and you said, how wonderful."

Well, what was there for me to do? I tendered my sincere apologies, explaining I was only trying to speak his language. Fine man that he was, he said that was ok. Little knowledge is indeed dangerous.

Lessons

My tenure was a difficult one. I ensured that the dignity of the office of the DA was not diluted and prevented the ambassadors from instituting practices contrary to the orders. It was this attitude that brought me into conflict with the ambassadors. It was during my tenure that the first visits of Chairman COSC and the Air Chief to Turkey were fructified in 2004 and a reciprocal visit of the Turkish CGS materialised as result in 2006, which we botched up thoroughly. My three annual reports were written, but none shown to me despite instructions to the contrary. Everyone knows what that means.

Negative experiences provide greater learning and thus I returned a much wiser man in every way. Of all the things, my appreciation for the contribution of the Army to the country at a pittance, in comparison to the MEA. I found all the three embassies that I worked with fell way short of expectations, despite being provided all the wherewithal by the government including financial support.

The Turkish Military were extraordinary hosts. Their warmth and hospitality made this tenure a pleasure. As for the MEA, they need to undertake an honest audit of the functioning of the embassies, both for what they deliver and how the foreign exchange is being spent. For example, they need to analyse whether the country needs embassies both in Syria and Lebanon or they can make do with one with accreditation to the other. Am sure there must be similar possibilities elsewhere in the world. It will save the country precious foreign exchange.

What must be done?

First and foremost, the Army must set the interse equation of the DAs right. You cannot equate a Col with a first secretary. The nature of responsibilities of a DA are far serious than a first secretary, who picks up this rank at around four to six years.

The MI Directorate must formulate a suitable course to be run by former DAs, who prepare the officers to tenant the appointment without anxiety and ignorance.

Must streamline the functioning of the DA to make him more independent. He should not have to float a minute sheet for a ream of paper, as sometimes happens in the embassies at the whims of the ambassador.

Make the DA invulnerable to errant ambassadors. The ambassador's report must be time bound, failing which, they must become invalid. The Superior

Reviewing Officers of the DA must review the reports, only after the ambassadors have written their part and do the needful rationalisation. It will help the officer maintain the dignity of uniform.

The DA's office should have a dedicated car. It will help him do a lot more without being at the mercy of the ambassador for every requirement.

Much after I returned to India, Mr Kuldip, the office superintendent, whom I had helped to get his wife's sizeable arrears reimbursed in India, wrote to me:

"Dear Sir, you will be happy to learn that I have been promoted, and with your blessings, I have been posted to the Indian Consulate in the USA——. Pl convey my regards to all Tom, Dick and Harry around you." He would be our representative in USA.

Embassies could participate in the TV serial "India has got Talent".

THE ROAD TO SUCCESS IN THE ARMY

Is there a magic formula that guarantees success in the Army?

When I was writing this book, one Brigadier suggested that I write a chapter on what it takes to succeed in the Army. I felt that it was not a meaningless question. He was hinting at something more than what meets the eye. I have attempted to answer that question in my perspective.

The Army is an extremely competitive profession. Unlike the civil services, where almost everyone reaches the additional secretary level and in rather quick time, the Army is entirely different. Almost seventy percent of the officers get superseded at the very first stage of selection, from Lt Col to Colonel. It can be very heartbreaking and disappointing because very competent, dedicated and excellent officers fall by the way side, simply because of the limited vacancies in the pyramidal hierarchy. It leaves them facing the stark reality of serving another two decades of marginalised existence where pride can take a hit.

So, what does it take to succeed in the army?

We need to start with the very basic question. How do you define success in the Army? Who is a successful officer?

Well, ranks are the most visible symbols of success in the Army, seen from miles. But, are ranks the only parameters of success? Are all the officers who never made it to the flag rank unsuccessful officers?

The purpose of growth in military career is to maximise your potential aligned with military mission and values as a leader, *in the interest of the organisation*. Anyone who starts with ranks as the goal is putting the cart before the horse. He will waste too much effort, energy and peace for too long, worrying over things he has no control over. Whether he gets there or not, he is certain to miss the fun of the journey of this profession, and the journey is always the bigger fun than the destination. By then, the best part of his life is behind him. I have seen many wrecked officers, including Major Generals whose lives seemed to have ended when they were not approved for promotion. One cadet who made his intentions of reaching at least one star rank clear, never went beyond Major, having had to quit in disgrace.

Contrast this with my caveats for command. I laughed and had fun every single day. Lived fearlessly and left a legacy in every command assignment. The last thing on my mind was my confidential report and promotion. That gave me the freedom to choose my response to any situation. One year after I retired, the rank and awards are forgotten and of no consequence, but I fondly remember the journey, trials and tribulations.

I know of an officer, who did reach very high ranks, but manipulated his senior reporting officer to become his colonel of the regiment. Is this carving your destiny? Is all this scheming, the intention of which does not get lost on your colleagues and subordinates, worth it? No. One should base his functioning on his competence, commitment and values and not on such manipulations.

Coming back to success. What about officers who reached the rank of Lt Generals and were court martialled? What about those Lt Generals who were decorated with a series of peacetime awards and were called for investigation by the CBI for corruption and their promotion orders cancelled? What about those officers who inspire no confidence in their men and who are happy to see them leave? What about those who reach the senior ranks, but their home front is in a mess? What about those, who have reached the highest ranks, but their period in the Army is seen as a dark period?

To my mind, in our profession, while they may not bring the same visibility and recognition, there are far more important parameters that define success than the ranks. Since, in the ultimate analysis, our profession deals with leadership of men in the extreme circumstances, and the competence to lead your command to succeed in battle, the major parameters of success lies in what an officer achieves in this field. Unless an officer achieves the trust and confidence of his command in his character, competence and courage to do what a leader is expected to do, he is a failure, never mind the rank one reaches. No one is ever remembered for the rank. One is remembered for how much one impacts the growth of the organisation and the people who were associated with him.

My definition of success in the military relates to professionalism, which I understand as:

Functionally:

- Service based on principles and military values.
- Passion and pride to make a positive difference to the organisation.
- Competence, intellectual depth and the vision to effectively respond to professional challenges and crises. It does not mean doing the routine things well. It includes the ability to train your command to accomplish the operational mission in the most cost-effective manner.

Personality wise:

- It means having the character and integrity to command the confidence and trust of subordinates.
- It means having the courage to stand by your decision and what is right irrespective of the consequences.
- It means leadership by personal example.
- It means selflessness and willingness to subordinate your interests for the organisation and your men.
- It means being good enough to be a role model for the subordinates.

In my perspective, in simple terms, if your officers and men are prepared to follow you, even when you have no official authority over them, you are a professional. Summarising the above, professionally, the success is determined by the positive impact that an officer makes on the organisation and the men. Unless he checks the above, in my view he is not successful in the military sense. An officer who has not lived by the military values and only lived for himself pushing his personal agenda and self-interest is a professional failure.

An individual can choose his own definition of success, but, for the success to be worthy of holding your head high in public, it must pass the 'value audit' of the end, the ways and the means in the context of military values.

It must command the approval of your colleagues and your subordinates, as having been well deserved and earned the right way. This is not to say that your colleagues will always be happy for your success, but will give you the due.

Your achievements should inspire your subordinates by the manner you have done so and emulate your principles and values.

Any success that fails to do this will please no one else. I have seen many army commanders and chiefs who cease to be anybody on retirement.

For any professional success worth striving for, it is a good idea to do it based on fair play and selflessness centred on principles and values. In a military career, if one has not been a role model for some who have served under him, in my opinion, he has achieved little of consequence. If you do not command a good spoken reputation after twenty years of service, to me your character and personal traits are suspect and you are a professional failure.

Having said that, sound growth leading to success is a function of how a person evolves himself at the following three levels:

Development of Self

An officer needs to build a solid foundation based on character, values and a positive attitude. He must base his life centred on principles and values. Without this, an officer will never achieve success that he can be proud of in public.

A man is the product of his thoughts. In the same era, there were Hitler and Gandhi, Saddam Hussein and Gaddafi and Nelson Mandela. Thoughts result in words. Words lead to actions. Actions form habits. Habits make character and character reaps destiny. On an average, fifty thousand thoughts cross a human's mind. Of these, if five percent thoughts were to be negative, which would be a massive two thousand five hundred, they could destroy a man. So, one needs to be watchful of his thoughts. One needs to surround himself with positive and right thinking people.

This development of inner being provides the foundation on which a person can build his career.

What does a personality based on character and integrity do? It enables an officer to serve without fear, head held high and look his superiors in the eye. It allows him to serve with pride and purpose. Above all, it enables a man to choose his response to circumstances and sleep without a worry in life.

Competence

Confidence comes from competence,
Competence comes from knowledge,
Knowledge comes from reading, observing and practicing.

Officers need to master the art of military living. This means developing himself on many fronts. He has to develop himself to be a good leader and be ahead of the Revolution in Military Affairs (RMA). He has to develop himself as an officer and a gentleman.

The conduct of warfare has changed more in the last sixty years than it has in the last six thousand. This is an era of invisible warfare. Information warfare, cyber warfare, unrestricted warfare, asymmetric warfare, fifth generation warfare, space warfare, etc. Technology will always be 'advantage terrorists'. Officers would constantly need to be abreast with technology, appreciate what cannot be seen, see around the corner and beyond the hill.

The fact that the terrorist are succeeding almost every time in surprise attacks in the Valley or in the red corridor where paramilitary forces are deployed, means the military leadership is not ahead of the events. It is reactive and not pro-active.

No military commander can be a great captain of war without being a student of military history. No officer can realise his full potential as a military leader without the knowledge of military history. What does this do? It enables a person to learn from past events and prevents repeating mistakes that will certainly cause loss of troops lives. Study of military history prevents repeating mistakes, at the strategic, operational and the tactical level, committed in the past. It helps to learn how and why some leaders succeeded whilst others failed.

It helps to open one's mind on the uncertainties of and possibilities in war.

Preparing for war is an expensive burdensome business
But, there is one thing that costs little-study.

— *Field Marshal Slim*

A reading habit, so essential to personal development, is conspicuous by its absence in officers. Where remaining current is a matter of survival, it happens through necessity and compulsion. The best results would be obtained if it happens by choice and through self-interest.

In the fast-changing world of technology, military leaders have to be voracious readers. The 'scholar warrior' term, much in use in the Army conveys the need for well-read officers. The senior military leadership needs to have a vast horizon and vision that goes beyond military knowledge. It covers politics, diplomacy, economics, governance, constitution, regional and global security understanding, bilateral relationships, etc.

There is no institutional methodology to acquire this knowledge. Thus, it has to come from personal reading.

There is one simple secret.

Reading one book a month, or devoting one hour every day on a subject of choice, beyond daily routine, enables a person to be within the top five percent experts in that subject in the world in five years.

I wish someone had told me this when I was a young officer.

Winston Churchill attributed his success in life to five years of voracious reading as a young British Officer, while in India.

Time Management

Time management and gainful utilisation of time are critical for an officer to develop himself on all fronts. A vast majority of officers are guilty of wasting their time and blaming it on the organisation. Far too much time is wasted on social media which does not contribute to one's growth in any significant manner. A simple reality check to assess what percentage of telephone calls is really necessary, will bring home the lesson of how much time is wasted on this one aspect. Add to this other social media tools, and the time wasted is colossal. Practically, Facebook, WhatsApp, etc., are social tools and have very limited knowledge value, which one can really do without. I have no doubts that they are colossal time wasters.

As commanders bulk of our time is spent on addressing the urgent, sometimes important but often otherwise, termed as fire-fighting. It must be remembered that the urgent never builds the combat capability, only the important does, which is never urgent. The bulk of the commander's time must be spent on this. The rest must be decentralised.

The Will to Kill: The Intensity of Desire

> *"Those who say they can and those who say they can't are both generally right."*

One often hears that hard work leads to success. That is only half the truth. There are many aspects that contribute to success. Unless all those are nurtured, hard work alone will not result in success. Often people talk of fate, luck, God's will and so on. Whatever that means, two things are clear. One is that of the above—except hard work nothing is in your hand. A person has to have a clear road map for his career and be clear that destiny and career can be charted by no one else but you. The hard work has to be focussed, correctly directed and persistent. One has to develop that attitude to see every failure as a stepping stone, for every failure teaches lessons and brings one a few steps closer to the goal. Realise that success lies on the other side of failure and persistence is the vehicle to take you across the line.

There are two voices of rejection that need to be thrown out. One is of the people who surround you, who will tell you why it cannot be done and the other, your own. Believe in your ability to persist and prevail. If you give up, failure is guaranteed. If you persist, success has more than a chance.

Two words that must be removed from your dictionary—excuse and blame. It is all within oneself.

Where does one start as a military officer?

The first thing is to start with the end in mind. How does an officer wish to be known when he hangs his uniform? What legacy does one want to leave behind?

Mission Statements

Every officer during his life time will shoulder many responsibilities, both on the professional front and the personal front. These responsibilities will keep changing, and mostly multiplying with the passage of time. To serve and live meaningfully, he needs to give a direction to his life on both fronts by setting the way points and identifying the goals in each role that he will play and how he plans to achieve them. That direction can be given by writing down a kind of a constitution for himself, which is called mission statement. The mission statement identifies many roles that an individual plays, and the goals he wishes to achieve in each role and how.

Framing a mission statement is a deliberate effort that needs deep introspection, meditation and deliberation based on the analysis of inner being, potential, own values and one's ambition.

The mission statement is a dynamic process that should be reviewed by the officer, every time his responsibilities change or increase.

Amongst all the aspects that are necessary for growth, the most important aspect is how badly one wants, what one wants. It is all about the hunger. One will achieve great things only when one has the passion, loves his work and is hungry. It is easy when you are crazy about it. It is difficult when you are lazy about it. In my scheme, in the Army, it is not the desire to achieve ranks, it is the desire to develop oneself as an inspirational professional soldier and a leader of substance that should be the aim of every officer. This should be the hunger that must drive an individual. Doing the other way around is certain recipe for missing the fun of the journey of soldiering and a likely disappointment at the end. The rest will follow.

What is needed to attain higher ranks in the Army?

I have deliberated on the basics that are necessary for a successful career that you can be proud of in a dark room, success according to my definition.

Sadly, all that does not guarantee attaining higher ranks in the Army. If anyone thinks hard work and competence alone will guarantee higher ranks, he is sadly mistaken.

I asked one senior army officer in 2017, "Does it need competence to do well in the Army?"

He replied, "Look at your last two chiefs and you have the answer."

Couple this assessment with what Shobha De wrote about the last two chiefs, and that says it all (page 201).

I then picked his brain on what was needed to be successful, if competence is not?

He said, "To succeed in the Army, purely in terms of ranks, an officer needs to be system-fit."

While it will be wrong to paint everyone with the same brush, in my perception, the Indian Army, while having a percentage of good Generals, as a general statement, lacks Generals of substance. This means that a majority are driven by careerism at the expense of the growth of the organisation or their subordinates. The majority will not accept dissent with grace. The majority have personal standards of morals and ethics, as it suits them, as such will be uncomfortable with an upright officer. Subordinates with moral courage, often are devoid of company. Probity and integrity could do much better.

What did he mean by being system-fit? Based on what I have seen, system-fit could include the following:

One—Developing the necessary skills and competence for a wide range of employability and doing the courses that enable you to punch all your cards to get all the weightages that the system allows, without which there is little chance beyond the rank of a colonel. Nomination to HC and National Defence College (NDC) courses are primarily a function of good reports. And good reports are less a function of competence, more a function of one's personal equation with the boss.

Two—It means adaptability, i.e. managing your environment and your boss, despite your differences of values and opinions with him.

This might mean closing your eyes to many untoward things which may involve the superior officer. All this will hurt the conscience of a righteous man. And this might mean swallowing your pride and values. As I said, an officer serves with a new superior almost every year and they could be as different as apples and oranges. Clearly, one needs adaptability of a different kind and sometimes, the ability to subdue the voice of one's conscience.

Haven't we had enough tainted Generals—booze General, meat General, dal General, Sukna General, Adarsh Generals, etc? They reached those ranks and hordes of officers served with them on their journey. Surely some officers played along, others managed the environment and some looked the other way and kept out of trouble.

When in the wrong place, system-fit means—Avoid-Adapt-Adopt.

Three—System-fit means, having an influential mentor who can help in critical times and do value addition to your career. It means selling yourself and ensuring that your boss knows what you are doing even if you are not doing it! I remember, my battalion was deployed in very active counter-terrorist operations in Sopore in 1991–93. One company would achieve nothing of consequence under a particular company commander, who came from outside. Yet, he would send a weekly achievement report of many pages to the commanding officer, located ten yards away. While the commanding officer understood, it did the officer no harm, I guess. I have narrated the incident of the commanding officer of the EME battalion, for whom I had placed my battalion out of bounds during Op Parakram, earning his spurs by wrong reporting.

In fact, the Indian Army in current form is not the place for brilliant people. Horse and a donkey progress at the same pace. Quite regularly the donkey outruns the horse by being system-fit and conformist.

Since, success in the Army is primarily a function of good reports, it is a place for a diligent average officer who can manage the environment and his bosses. Someone who can subdue the voice of his conscience and live with it. Some righteous officers do make progress; chances are they learnt the tact to adapt, avoid and manage the environment when they were in the wrong place with a wrong man.

I know a commanding officer who told his second-in-command, "You look after the regiment, I will look after the commander." For sure, he knew how to be system-fit.

The next success secret (system-fit trait) lies in the art of mastering the powerpoint presentation and making old wine look new in a new bottle. Command assignments occupy only a small space in the long career, no more than six years. Thus, in the second half of his career, an officer is assessed on non-combat or non-leadership type of assignments. Mastery in powerpoint presentation and communication skills makes an officer indispensable to his superiors and thus system-fit. Develop your communication skills and you have the tools to be system-fit.

System-fit means being well versed with social graces (read street smartness). Do not lag behind your competing colleagues.

I was narrated this interesting incident by the officer who replaced me in Turkey as the defence attaché.

"When you go to the next assignment, present a Black Label to the Brigadier General Staff (BGS), he is your initiating officer, the Chief of Staff (COS), he is your reviewing officer and the Corps Commander, he is your senior reviewing officer (SRO). Wait...wait, don't waste the bottle on Corps Commander, as he only gives one box grading and he will endorse what the rest have given."

I said, "Buddy, I have seen this Army three years more than you. I never gave a gift to anyone ever."

He said, "This is what happens in the armoured formations. You see, when I was commanding my battalion in the armoured brigade in Patiala, the commander held an open house on his anniversary, not that he was paying for it. I went with my wife with a bottle of wine and a bouquet. At the same time, the commanding officer of the armoured regiment also landed there. He was carrying a bouquet only. Seeing him, I felt comfortable that my gifts were better than his.

During the entire party, the commander and his wife were very warm to the armoured regiment commanding officer and his wife and very stand-offish to me and my wife. I kept wondering why? I thought to myself that this guy only brought a bouquet, I brought a bouquet and good wine, why then?

I quietly found out what this man had presented to the commander on his anniversary. My hunch was right. He had presented a silver tea set, which was delivered well before the guests arrived, and himself landed with a bouquet in public view.

I was not to be left behind; the next day I presented him a matching silver salver, with an excuse as to how it had got delayed. I was back in contention and all the animosity of the commander and his wife had evaporated."

All three officers did well in terms of ranks, the commander and the two commanding officers. This is not an isolated incident.

Since promotions are mainly a function of reports, system-fit, at the bottom line, simply means being in a great relationship with your bosses. Very few and very rarely, have the officers been sacked for incompetence. They have often been for not getting along with their bosses, despite professional brilliance.

This is not to say that all commanders fall in this category. You will find many who will appreciate your candour and righteousness, as long as you do not step on their toes.

Well, with all this moral lecture, what did I do in my career?

Well, I gave no gift to any one in my career, as a commanding officer or commander, none whatsoever. As a divisional commander, only bouquets were presented to the corps commander's and the army commander's wives when they visited my division. And I paid for those bouquets and not the division. I presented these commanders with my book on military humour and nothing else. No gift was given to any commander of any rank who came to the division to attend the division operational discussion.

For the wedding anniversary of anyone from a young captain in the division to the army commander, the AWWA card was sent and nothing else.

In my division, gifts were prohibited for anyone who was invited to my house for dinner or any other reason. It was categorically mentioned on the invitation card, '*Strictly, no gifts please*'. The staff officer who received the guests politely advised them not to take out the gifts, if they had brought one despite the advisory. On the very first occasion, I returned the gifts of my deputy GOC and Colonel Q, right outside my house. It never happened again.

There were no gifts of any kind on relinquishing battalion, brigade or division command, except my own headquarter memento, even though some formations and units did send their mementoes after I had relinquished the command and left the station. I kept these mementos with pride while in service and returned them to the units on retirement with a letter of thanks.

Well, I could appreciate their dilemma. They had rarely come across a commander who meant it. They were not quite sure what to do and would err on the positive side, as they saw it.

One commander confessed his predicament. For the finals of the inter-formation sports competition, there were no sofas or shamianas for seating;

only chairs were arranged on a tarpaulin for me and other spectators that included ladies. I sat down, not even noticing that. The commander very guardedly asked me, if this arrangement was fine, as per my directions.

I said, it was great. In fact, I said, you have good grass and shade here and not even the tarpaulin was required.

He said that it had taken one week's deliberation to do this: "We have never seen such a thing and we were wondering what to do, as we were not sure whether you meant it."

I understood their dilemma. The next time a unit faltered, I confiscated the red runners and carpets of a unit to drive home the point that I meant what I said. Later, that commander mentioned to me, they were back in dilemma when the new divisional commander took over. Everyone went back to red carpets and runners; to be on the safe side.

So, every individual will set his own standards of success, notwithstanding my views. But, at the end of the long service, if he passes the audit of 'the end, the ways and the means' that are compatible with the military ethos and values, and his colleagues are convinced about it, he can be proud of his achievements.

Soldier and the Home Front: Managing the Conflict

Amongst many things, one thing of great significance that happens to an officer on commissioning is that he assumes an unlimited liability in service of the nation. What this effectively means is that he writes a blank cheque in the name of his country. This seemingly innocuous condition plays a major role, when the professional and personal interests clash, as they are bound to happen sometime in this profession.

The outcome of this clash has the potential to create friction in the family. As the divisional commander in Meerut, one infantry officer narrated that he could not be present for the funeral of his six-month-old son, who had died when he was posted in the field.

Let us for a moment look from the perspective of his wife. There was no war on, even though he was deployed in a remote place on the border. The reason for his absence at this monumental tragic time will never be of any significance to the mother. No reason will ever justify his absence to his wife. This officer will always be on the backfoot at home, for life.

My personal experience. I was not present when both my sons were born. This is an event when the wife needs the husband around for a hundred things, and a hundred things can go wrong. If they do, and he was not present, the father will live with guilt that can be devastating. No one else can replace him at this time.

I was on a month's leave after the staff college. During the road journey, my one-year-old son fell sick and had to be admitted in the hospital. One month into the hospital, he was being given seven injections a day for what eventually turned out to be a wrong diagnosis. Despite being apprised of the situation, the commanding officer refused to extend my balance of leave. The unit was in the Poonch sector, but there was no war on. I had been out of the unit for two years and they were fine without me. But the commanding officer decided that I cannot be given any extension of leave, even in this circumstance. I had to take my son out of the hospital in that condition and leave my wife to fend for herself with two kids both under two and half years in age. The commanding officer was off his rockers to behave in this manner. How does he expect to win the officer's commitment and loyalty?

Many officers have similar experiences. What is the housewife expected to make of this? While they understand the peculiar service conditions, they will find it hard to be convinced, for this places tremendous challenges on them, particularly when they have no support.

On the other hand, our Army seems to have mastered the art of keeping unnecessarily busy. The cutthroat competition means that the officers who are in the proverbial 'rat race' or too ambitious will often fail to balance their professional and personal obligations. In some cases, the nature of their appointment and locations do keep them busy, but don't we have their deputies who should hold the fort in their absence. Very few commanders or divisional commanders allow their principal staff officers to go on longish leave or even career courses. It is an indicator of their own incompetence. I was commanding a Rashtriya Rifles Sector in Lolab Kupwara, the hotbed and the most challenging insurgency-hit region in the Valley. The GSO (1) came up for promotion. I was told by the Army Headquarters that I could take up a case for holding him back for three months, because his relief cannot be posted before this time. The officer himself, Col Balbir Singh, 4 GR, offered to stay on. I sent him and the G 2 ran the show. We managed everything without fall of efficiency. It only meant more involvement of the deputy commander and my own. One Colonel GS told me that he had availed only three days leave in his entire tenure of two and a half years.

Sometimes, the officers in honest commitment or a sense of responsibility join their units in field when they had the opportunity to be in peace with their families, just to be part of the unit in action. A well-intentioned initiative, unless the wife has the support and the safety to manage alone, is bound to create a life time complaint that will come to haunt an officer.

The officers have to realise that running the house is a real challenge and doing so alone is tough, both when the children are small, as also when the children are grown up. In the latter case, the lifestyle of the current teenage generation invariably is in conflict with the beliefs of the parents.

In retrospect, I have a word of advice for the highly ambitious kind. The Army is not standing on your shoulders alone. No one is indispensable. It is an institution that is designed to function in absence of the top incumbent. If you have not trained your number two to step into your shoes and function, you are a failure.

While it is important to do well in one's career, one must remember the following:

- Life is a wholesome package. Lack of balance where one has a lop-sided growth in one field at the cost of others is a recipe for long term frustration. In the pie chart of life, many facets make up this package. The lasting source of happiness will come from what happens inside your home and the biggest stresses too will come from the same source.
- Happy domestic environment will help the officer succeed in his career. This domestic happiness will come, not from your growing ranks, but from your playing your part as a husband, father and son. It is possible to do your job seriously and find time for your personal obligations.
- One week from your grave, your rank is the last thing that will be on your mind.

Defining Success

At a personal level, how does one define success? I have laid out my views on this earlier in the book. It must be achieved keeping balance and congruence in all aspects of life.

Why are there so many anecdotes on wives and husbands? Why do both feel that one does not understand the other? Sadly, even in the Army, as in the civil, cases of matrimonial disharmony are on the increase. Is there a way to stem the tide? The ladies have to realise that the Army, while empathetic and supportive, does not afford the liberty to officers to cherry pick their absence from the unit. While the senior officers attempt to help their officers in need and critical times, there may be occasions when the ladies will have to manage their affairs themselves. This in particular is applicable to officers in the hot seats. Yet, officers in such appointments also put self-imposed constraints on themselves when they may not be required. I remember a corps commander telling me that he had availed full leave every year of his service. That is the point I am making for junior and senior officers. The Army will function without anyone. Maharaj Sadguru says, "When you start thinking that you are doing something very great or important, step back and take a holiday." Meaning, the organisation will run even without you and possibly better.

Stephen Covey in his famous book *Seven Habits of Highly Effective People* gives a simple solution to resolve interpersonal misunderstanding. It is important that each partner first attempts to understand the other, before he expects to be understood. When one understands the other's perspective, it becomes very easy to find a way forward.

He narrates his own experience, where his wife insisted on buying a particular brand of refrigerator, which meant it had to be bought from a store located long way off, needing a whole day and high transportation expenses. The husband insisted that they buy an equally good brand available next door which was very convenient and avoided the transportation cost. This led to conflict. Finally, when Stephen asked her why she insisted on this particular brand, she narrated her story. When she was a kid, her family slipped into a major financial crisis that threatened their very survival and dignity. At this time, when the company was shedding employees, this company came to their aid. It not only kept her father on the job but also gave them financial assistance to get over the crisis. It was her emotional debt to the company that she wanted to repay by having their fridge in the house. Suddenly, her point of view made sense and all disagreements disappeared.

Since marriage is about two individuals coming in to spend life together, they will have different interests and concerns. At times, marriage may bring two persons to live in the same house with entirely different natures and personalities. What the couple needs to work out is a methodology that one does not impose himself on the other. The environment must provide space to both to pursue their interests, if both cannot bring themselves to do them together. At the same time, it is important that there must be an effort to do maximum things that can be done together as a family. This would need accommodation by both the partners.

I have often heard stories of officers and men who get no support from their spouse in looking after their parents. That becomes a sore point in the married life and can cause tremendous stress. The solution lies in an early pre-emptive strike. I was narrated this incident by one officer. He said that very early in his marriage, he had decided that the first cup of tea of the day on wintery cold mornings will be made by him and served to his wife in bed. His wife did the rest during the day. I told him, this is the best master stroke I have heard. He, by initiating and accepting a tactical defeat had earned himself a strategic victory with far-reaching dividends. His wife would often tell him, you have done enough in life, it is for him to relax and for her to work. It is not because of that one cup of tea, but his willingness to share the workload at home voluntarily. He did so before his wife even asked for it.

What needs to be understood by both the partners is that marital bliss is impossible to attain unless the concerns of both partners are addressed. While there is the immediate family, equally important, there is the extended family that includes the parents of both partners. Parents are dear to both. The parents will need different kinds of support at different stages of life. It may be in the form of financial support, taking care of siblings, and periodical visits to them in the early stage, medical support and keeping them with you in old age, and above all, the important emotional aspect that you care. Caring of parents needs the whole-hearted cooperation of both partners because it places demands on your time and resources. While it is easy to support with finances, it is demanding to provide personal time and care.

What does one need to do? Early in the married life, the parental responsibilities do not come up as the parents are young. As Stephen Covey suggests, the couple needs to write a family mission statement or a constitution which will guide their lives as a family, with due deliberation. It must spell out the aim, which is marital happiness and family growth. It must identify issues that are dear, important and sensitive to both. This will identify areas of conflict that must be avoided. It must spell out how they plan to achieve the goals in every role that they play as a military couple, as individual parents, as sons and daughters, as children-in-laws and citizens.

It is only natural that own parents are dearer to each than the partner's. It has been a historical fact that the son-in-law rarely has problems with his in laws. That is not true for the daughter-in-law. It must be understood, that it is well-nigh impossible for a man to discharge his obligations towards his parents without the support of his wife. A man must think ahead and take initiative to look after his partner's concerns and her parents before he can look after his own. And this must come as his own initiative before the spouse asks for it. Do it and you will earn the tacit cooperation of your wife in discharging your duties to your parents.

This need will become greater as they get older and would need to stay with their children at some stage in life. Invest early and you will reap domestic happiness. Amongst many key areas, which dictate marital happiness, this is one of the most important.

When I look back at my life, I realise that I earned myself a good name on all counts in professional and personal life because I had centred my life on principles and selfless service, often sacrificing my personal interests for the organisation. From the home perspective, I was found wanting as a husband and a father, which led to her facing all the challenges of running the house alone with two small children. When I see from that perspective, I think she is right. The husband has to be around when needed at home. I was not.

These requirements are when children are small or the wife is unwell, children's education needs, holidays, participating in important home functions, being around during transfers and difficult times, etc. If these are not addressed, friction is the natural consequence and your success in professional life will mean little to your family and it brings no pleasure.

The Army by itself cares little for what you sacrifice for the organisation. There are very large number of officers who rose to be Generals who have not seen a shot fired in anger. Of the commanding officers who commanded the battalions in the Kargil war, only one reached the rank of Lt General. A soldier must realise that while the nature of service demands that he places his professional duties first, he must find ways and means to ensure that his obligations as a husband, parent and son are addressed too. This will enable him to perform his duties well and find happiness at home. The superior officers need to play their part. They must demand the presence of their subordinates with due consideration. They must remember, if his heart is not in it, there is no point in having him in the office.

Disagreements between husband and wife are common place and quite often they can be unpleasant. Both partners need to make a deliberate effort to adjust and accommodate each other. When there is an issue of conflict, both partners are better advised to limit your criticism and arguments to the issue at hand. Attack the issue and not the partner, which can be the case. Invariably, one argument leads the couple back to past issues, some more than two decades old that have no relevance to the issue being argued now. This badly harms mutual empathy that keeps simmering underneath and surfaces at the least provocation. This leads to situations which can ruin the home environment for days. As long as a partner remains 'historical', a tendency more likely with the wives, and keeps digging issues out, that should have been long discarded and buried in the dustbin of times, they will continue to fester friction. This does the family no good. It ruins your future because of the past, no matter who was responsible, which itself, in case of a couple, will never be resolved.

Thus, the ability to forget and forgive the past is extremely important for marital bliss.

Finally, there will be arguments, often on inconsequential and silly issues, spiralling into accusations and counter-accusations. Discussing marital harmony and disharmony amongst a group of friends, one lady made a brilliant point. She said, many times we have our differences, but one thing I am very clear.

I am not prepared to win an argument and lose my husband.

Great Words of Wisdom.

Tongue is a deadly weapon in marriage. Nothing can be sharper and more destructive. The couple needs to remember that in proving the other person wrong and winning arguments, beware, you may be losing a bit of your partner's empathy and concern for you bit by bit.

Remember, amongst couples there is never a win–lose or a lose–win situation. If it is not win–win, it is a lose–lose situation. If you remember these words, you will make efforts to make sure, it is win–win. If both show maturity and wisdom, leave their egos outside and have the sense to appreciate what is good for the family, I believe it is possible. It is worth the sacrifices that both make for a common cause—the marital bliss.

The army officers or for that matter service officers suffer from a peculiar circumstance. They do not have the benefit of anonymity as do the civilians. In the Army particularly, where there is a soldier buddy at home, all occurrences at home become public knowledge in the unit in no time.

Work and Life Balance

What is it about?

I said that it is important for an officer to get the balance right. How does one do it? This is the challenge that officers who pursue their career with the aim of reaching higher ranks will need to contend with. In the army, all commanders have a short tenure. They have to make a mark in that short tenure and his top priority is likely to be his own interests. He wouldn't mind showing concern for his subordinates' 'work–life balance' as long as his interests are not affected. Most commanders will be willing to accommodate your work–life challenges if you have earned it with performance, without affecting his.

The general belief is that family obligations are an individual responsibility and the officer must find his own ways of handling it. More, that the army wives are expected to handle lots of issues themselves. In our kind of a career, you cannot argue against that belief. For, when an officer is in field, the housewife manages most of it herself with little or no help from the man. Today, with working ladies, the demand on the officers to contribute will only rise in future. A lady has to make a choice, as would a working woman in a corporate. It will be a trade-off between working and home. A career officer has little control over when he can be available and when not. While no ladies complaint about being separated when the husband goes on a UN mission for obvious reasons, they complain about not being sent on leave when in India.

It is pretty simple, while officers who have accepted their fate and given up on their career, maintaining work–life balance is rarely a problem. Not much is expected from them by their commanders. For the running horses, they will have to find their own ways of striking the balance. It will need good understanding and accommodation of their spouses.

Lessons

It is most important to get the balance right in life. Make sure you get your domestic front in order. This means marital harmony, children's education and future, family health, financial independence and meeting the obligations of other roles that an individual performs in life. Without success here, no achievement is worth having. Your greatest happiness and worst stresses will come from what happens in your house. While an officer may be the philanthropist kind and deeply rooted in military ethos and values, the housewife is focussed on the family. Her belief in the military ethos may not be so. While that may sound too narrow focussed and selfish, it is in the best interest of the house, in her perspective. Do listen to her and take important decisions based on mutual consensus.

Do write your mission statements as a husband, as wife and as a military couple. Those mission statements, a product of deep introspection and thought, values and meditation must specify your goals in all the roles you play in life and how you wish to achieve them.

This mission statement of a housewife shows how she plans to discharge her varied roles.

"I will seek to balance career and family as best as I can since both are important to me.

My home will be a place where I and my family, friends and guests find joy, comfort, peace and happiness. Still, I will seek to create a clean and orderly environment, yet liveable and comfortable.

I will seek to exercise wisdom in what we choose to eat, see and do at home.

I specially want to teach my children to love, to learn and to laugh—and to work and develop their unique talents.

I value the rights, freedoms, and responsibilities of our democratic society. I will be a concerned and informed citizen and exercise my right to vote...

Will be a self-starting individual who exercises initiative in accomplishing my life's goal.

I will always try to keep myself from addictive and destructive habits.... I will develop those habits that help me expand my capabilities and choices.

My money will be my servant and not my master. My wants will be subject to my needs and my means."

(Stephen Covey, *Seven Habits of Highly Effective People*)

A lot has been written on the frailties of this institution of marriage. In the West, there is the concept of love marriage. Apparently, the couples know each other and understand each other before they get married. It does not work in high percentages of the marriages. In India, we have the concept of arranged marriage. That also does not work in an increasingly larger number of cases, but the couple keep chugging along in misery, for children's sake or because of social reasons.

So, the question for the youngsters is, 'Is there a way to find a compatible partner where life will run smoothly?'

I guess if there was one, people would have found it. I am not a believer in horoscopes and 'Jyotishis'. But it may not be a bad idea to follow the traditional practice of matching horoscopes, for one simple reason that two individuals make the most critical decisions of their lives. Today, that seems to be the only way, whether effective or not, that the Hindu society has faith in to assess the compatibility of two individuals for lifelong companionship. If nothing else, it gives you mental peace that you did whatever was possible.

I believe two mature individuals can live in harmony by showing necessary maturity and compassion, despite different personalities.

We have seen possible causes that lead to friction. Based on those and peculiarities of the military service, I have compiled a set of questions of my own for an army bridegroom.

Important questions a young officer should consider asking the girl whom he intends to marry.

1. Do you understand the challenges of marrying an army officer?
2. The nature of my profession entails that I cannot always be at home when you would want me to be. You will need to take care of bulk of the responsibilities, most of my career. Are you okay with that?
3. How important is it for you that I attain higher ranks? What are you prepared to sacrifice for it?
4. I am an extrovert/flamboyant kind of person who is noisy and loves to laugh and talk to people/introvert. What is your take on my nature? Are you okay or would you want me to change?

5. I have few exs. What kind of relationship with them will you be comfortable /uncomfortable with?
6. We are a close-knit family. I deeply regard my parents. What support can I expect from you or do you have a problem with that?
7. What do you expect from my parents in terms of assistance and support, etc.?
8. I have siblings and responsibilities for which I will need to support them in different ways. Do you have a problem with that?
9. How do you want me to conduct with your parents and siblings?
10. I love my space where I love to do things that you may or may not be interested in. Do you have a problem with that?
11. Every marriage has quarrels and disagreements. What do you think is the best way to resolve the disputes and disharmony?
12. How is the environment in your home? How are the disagreements resolved?
13. What in your view can lead to serious differences that threaten marital harmony or marriage itself?
14. Where do you see the marriage ten years from now?

The officer on the other hand must answer these questions himself.

1. Do you understand what responsibilities will fall upon you when you get married?
2. Are you up to it to make the sacrifices and discipline yourself to make success of marriage?
3. Do you understand the nature of the girl you are getting married to and her expectations?
4. Do you understand that marriage is about accommodating each other's differences and enjoying the similarities?

Hurry Curry. Time is always critical in life. But, there is one place, where there should never be a hurry to decide under the pressure of circumstances or a time deadline—marriage. Start early, take your time to find a compatible partner. If you are not convinced, it can wait. Do not jump into a blind well. The fall can be very painful.

The above process does not guarantee that you will find a compatible partner. Maybe, you will better understand what lies ahead. And forewarned is forearmed—it will help you to accommodate, adapt or adjust and avoid each other's red lines. Someone rightly said, if the army wanted you to have a wife, it would have issued you one.

I suppose, most wives feel, that there is only one perfect husband, and every neighbour has it.

Ghorakhal: Shangri La

At School, 1975. Flanked by SK Yadav and Late, Col JKS Rana

SURMOUNTING THE ODDS AT THE IMA, DEHRADUN

Receiving the Sword of Honour, June 1979

The Gold Medal

COMMAND WAS HUNDRED PERCENT WORK - HUNDRED PERCENT FUN

Bringing Luck; Gun shot through the helmet

A formation mobilization

We did not mind getting wet and cold during training

Never a dull moment with the troops

In a Jolly Mood

Office time is beer time: Officers Rugby

RUGBYING 10 SIKHLI BACK TO SHAPE

ALL WORK NO FUN MAKES BHOOP SINGH A DULL SOLDIER

Young officers on a fun evening at the Flag Staff House in Meerut

No free drinks: Rope trick at the RAM House

WHY SHOULD MEN HAVE ALL THE FUN

Division children football championship

CO 28 PUNJAB wins the prize for regimentation: Spouse's regimental colour jewellery

PART 2
MY PERSPECTIVE:
WHAT AILS THE INDIAN ARMY?

While this book is more about my experiences, I felt it necessary to record my critical perspectives on important issues that are eroding the very foundation of this fine institution. As I stated earlier, my criticism in no way diminishes the value that this great institution brings to this country. It will always be a pipe dream for any other institution to live with even a fraction of the commitment and accomplishments of this fine organisation, notwithstanding the continued degradation of this fine institution by the politico-bureaucratic combine.

The case where one Army Service Corps (ASC) officer knocked the doors of the court, questioning 'why should he be posted to operational area when the Army has declared him non-operational', is indeed a sad story on the vision and maturity of the senior hierarchy.

My criticism is meant to highlight the issues that need to be addressed for the long term good of the Army and restore that all important trust between various components of the Army through fair play and transparency. The Army has its own high standards, and those standards must be followed by one and all. I have raised most of these issues during the course of my service to the highest Army authorities and seen little happen. I hope, the powers that be will take note and stem the decline, for we have our own creed and ethos and there can be no other way.

Army Promotion Policy and Practice: Promoting Mediocrity

Hav Ishar Singh was one of the twenty-two heroes of the battle of Saragarhi in 1897. A lion-hearted man, he was described by a British officer as, 'nuisance in peace but magnificent in war.'

What would be the fate of an army, if it promoted officers who were otherwise?

The promotion policy and practices of the Indian army is a vast subject which repeatedly comes into sharp focus as officers take recourse to the courts for redressal. The subject is deserving of a thesis, as such cannot be dealt with in detail here. I wish to point out that the highest discontentment against ACRs, vacancies and promotions remains in the Army of all the three services. It begs the question, why is it so? The sole purpose here is to highlight major issues in this regard.

The past few years have seen serious discontentment in the Army on promotion aspects. Officers of all ranks have gone to the court, not satisfied by the fairness of the promotion policy dynamics and redressal system of the Army which has lacked consistency, depending on the Chief at the time. On two occasions in 2015–2016, two Major Generals, who were the only one to be considered (called agenda) in the promotion boards to Lt General for the lone vacancy, finally went to the court, seeing that their result was not being declassified for too long, possibly waiting for them to retire, and got their promotion. In number of such cases the court pended the retirement orders or notionally reinstated the officer with promotion indicating that all was not well with the procedure being followed by the Army. When I mentioned this to the Director General of the affected officer, he replied, these days, the only way to get your promotion is to go to the court.

I have two observations on this comment. One, that coming from such a senior officer, it demonstrates a lack of confidence in the system. This is serious. Two, that he being in that senior position, and seeing his own officer being the victim, must not sit on his haunches and take up a case appropriately as is his command function.

Why is there such discontentment? Reasons are many; some outside army's control, but of their own making which they are attempting to redress. This relates to the high ratio of regular cadre officers to short service commissioned officers. This means that, given the limited vacancies, large number of officers are considered for promotion but only a fixed percentage can be promoted. Sixty-five percent officers get superseded at the very first stage at the age of 36 to 38 years and are left to vegetate for another twenty years with a loss of self-esteem. Why has it come to this state? Clearly, the policymakers in the Adjutant General's Branch and the MS Branch for ages have lacked the long-term perspective of the problem, as such one change creates another problem and the need is felt to review the same. And, ironically, presumably the officers with the best record of service are posted in the MS Branch.

There is an unstated inter-arm rivalry between the infantry and the mechanised forces, comprising the armoured corps and the mechanised infantry. It has emerged as a consequence of pro-rata promotions seen by the mechanised forces, as a deliberate attempt by the Infantry to suppress them for fear of losing promotions in higher ranks.

The status of mechanised infantry has been an issue of repeated debates. The world over, there is no separate mechanised infantry arm. It is only in India that the infantry is trying to invent the unnecessary with a parochial mindset and getting away with it because they man all the important policymaking appointments. And the Chiefs, also from infantry, have been parochially dithering and silent. Reason—fear of losing vacancies in senior ranks. Pound-for-pound a mechanised forces officer has a wider professional horizon, is more versatile and confident and innovative. There is no government order designating mechanised infantry as a separate arm, even though this regiment was raised in 1979. Infantry, quite amusingly, is making deliberate attempt to keep the mechanised infantry out of the' infantry arm' in a rather unsoldierly and comical way. In 2011–12, the regimental flags of The Guards and Mechanised Infantry Regiments were not displayed at the Infantry Day wreath laying ceremony at the Amar Jawan Jyoti and their Colonels of the Regiment not invited for the wreath-laying ceremony. The Colonel of The Guards, then an army commander, came down heavily on the Director General Infantry to restore order. I was on a visit to Infantry School, Mhow, in Oct 2016. I was surprised to see that the regimental flags of these two regiments were also removed from the Infantry Mess without any communication to their colonels of the regiments. This is not trivial. If all this is happening without the knowledge of the chiefs at the time, then it is a shame. If it is happening with his consent, then he should be a man enough to do it formally.

The pattern of promotions to Lt Generals from 2013 to 2016 and in 2017 left a lot of resentment in the environment for good reasons. It is incredible that out of seventeen officers of the mechanised forces i.e. the armoured corps and the mechanised infantry of the 1980 and 1981 batch, only one was approved for promotion to the rank of Lt General. This, when this arm is already subjected to strict filtering through pro-rata vacancies in HC Course, in Brigadiers vacancies and the NDC Vacancies. It was perceived for justifiable reasons that the system including the Chiefs were not being fair in reporting on Maj Generals of the mechanised forces who were being systematically sidelined.

The A.V. Singh committee report, instituted after the Kargil war, forms an important input in the promotion policy in vogue today. The skewed interpretation of the A.V. Singh committee report by the Army in favour of the infantry, is the root of all the discontentment today which has been widely contested within the army. Calculation of colonels' vacancies based on the duration of unit command tenures in various arms and services is an intentional parochial flaw by the infantry. The duration of unit tenures of various arms and services used as the basis for calculation of colonels vacancies is not even being followed, as such irrelevant and untenable.

This becomes the basis to calculate HC, Brigadiers and the NDC vacancies on the pro-rata basis, which kills merit at the expense of organisational effectiveness.

There are frequent representations against the promotion policy, but practically little gets done. So often, I have seen letters by the army commanders to the chiefs put in a loop, with a remark to get the views of some dealing branches or ordering a study with no follow up. Ordering a study, while theoretically the right way, becomes a fig leaf to bury the issue in a soft kill mode. In all such studies, the most critical studies are headed by an infantry officer for obvious reasons. In due course the person who takes up the issue and the Chief both retire, killing the effort. The cycle keeps getting repeated.

Gen Bikram Singh had started a good practice of the retiring officers' tea with the Chief every month, during which all pension documents were presented to the officer, which saved the officers a lot of harassment. The next Chief, General Dalbir, introduced another good practice; five officers of different ranks would speak for two minutes each with recommendations to improve the system.

On my turn, I was nominated and I made these points:

- The relevance of current ACR document (annual performance report) to assess the character trait of a star rank officer. In my opinion it is a defunct document. The Army has been embarrassed by Generals and not by middle rung or junior officers, being in the news for the wrong reasons, whether it is meat General, dal General, booze General, Sukna land scam, Adarsh scam, or Augusta Westland case etc.—the list is long. If you were to study their ACRs for the last two decades, you will find that they have been consistently graded outstanding '9' in integrity in all their reports without exception by about 50 to 60 reporting officers in a three-tier reporting chain. Why is it that these perfect 'integrity' stalwarts turn out to be an embarrassment to the Army purely because of lack of integrity? The reasons are known to everyone.
- These officers did not become corrupt on becoming Generals, they were always corrupt and many of their superiors would have known that. But it takes a straight man with courage to act against a corrupt subordinate and the Army does not have too many Generals who can take action against corrupt subordinates fearlessly. Perks and privileges are a fig leaf to misappropriate/misuse resources and that becomes their vulnerability when it is time to take action against a corrupt subordinate commander.
- While selecting an officer for the post of defence attaché as a colonel, the MS Branch uses ACR only for the purpose of shortlisting the officers. It then seeks additional inputs on very important aspects relating to character traits, integrity, selfishness and suitability of the spouse etc, from their last two reporting officers. But it does no such thing even when selecting Lt Generals for promotion. Who is more important, a colonel defence attaché or a Lt General who could be the chief one day?
- The second point I made and that hurts me the most is the eroding relevance of the Army voice, and that means the senior military

hierarchy, in the MOD and the governance, even where it relates to national security and matters military. It is my impression that over the years, the erosion of our voice is so acute that the army commanders feel that they are just another man. One army commander said exactly that. That is alarming to say the least. How can the army commanders, just seven of them, feel helpless and incompetent to be heard in the ministry and the government? Why that has happened is because over the years, senior army officers have been bootlicking even the junior and middle rung MOD staff for petty favours such as postings, government clearances for foreign trips, unauthorised layovers and such petty insignificant favours. One needs to undertake an audit of the unauthorised military vehicles given to paltry bureaucrats in the MOD for petty favours.

- Is it not a shame that Lt Generals, and sometimes even the Chiefs, keep waiting till the eleventh hour for a government letter for a foreign trip, where it is the approved foreign travel plan with a sanctioned budget. And these senior officers are ready to run around like a rabbit to catch the flight. It is my opinion, the Army must demand that it be treated with respect, and except where the political, diplomatic or security circumstances have changed, no cancellation be accepted. Just imagine, what impression we leave on foreign hosts, whom you cannot confirm even two days before the visit. I have experienced this as the defence attaché in Turkey. A General officer worth the name must state that if he did not get the clearance so much in advance, he will not undertake the visit. But sadly, a foreign visit is more important to most than their pride.

The third point, which I chose not to speak in the gathering because of the presence of the MOD officials, was about the relevance of the army commanders and the personal staff officers (PSOs) to make a difference to the Army, within and outside in the corridors of the MOD and the government. When I made this point to the then Vice-Chief with examples, he said, 'I hope someone does something about it.' Again, a Vice-Chief saying the same thing! Who will bell the cat?

We understand that each level of rank brings some perks and privileges. These are meant to help you do your job better. But the Army is the bigger culprit amongst all the services in misuse of manpower and resources. Retaining sahayaks well after retirement, unaccounted manpower by General officers in Delhi, fleets of vehicles by whoever can manage is the common practice.

I spoke to a coursemate who had taken over as the Air Officer Commanding in Chief, Western Air Command. I asked him how much money he had spent on doing up the Flag Staff House before he moved in? He said, he needed to get some almirahs made in the house. Including that and white washing, he spent Rs two lakh from the MES funds after taking clearance of the Chief's office.

And how much from the regimental funds? I asked. He said, zero.

My Army friends need to learn a few lessons. The senior hierarchy in general is more inclined to leave their legacy in creating five-star guest rooms, renovating their flag staff houses and golf huts than in professional arena.

I have quoted two Generals, one from the Ordnance Corps and the other from the Army Service Corps who were being investigated for corruption. I bet they would have been graded 9 in integrity in last twenty ACRs. Their reputation is well known in the environment and there we are; they are highly decorated for distinguished service and approved for promotion to Lt General. One of them repeatedly kept getting posted in Eastern Command against all norms. This is not lost on the environment.

Fault lines in the selection procedure of General officers

The army is like fish; it rots from the head. We have to ensure that the head is forged in the crucible of character and integrity. There are three major traits that a senior officer must possess: character, competence and courage. The present promotion system does not address this important aspect. The Army has periodically refined the promotion policy to iron out the shortcomings. But, it is the quality of people who comprise a system that determines the success of any venture. The quantification system was meant to overcome the shortcomings of the previous policy in which decades of effort of an officer could be wiped out by one reporting officer. The value judgement of five percent marks given to the board, a good concept, lost value because of biased implementation. This also has serious shortcomings that need to be addressed.

I am aware of very honest efforts by officers like Lt Gen K.H. Singh, Infantry, to put in place a fair pro-rata and quantification system. In one instance, he even endorsed on file, "God save the Indian Army, if there is a war in next five years." Unfortunately, unless the majority live the ethical way, and rise above narrow arm/regimental parochialism, the Army will be the loser. And when the day of reckoning comes, men will pay the price and the country its honour. Similarly, Maj Gen Vinod Vashisht, Artillery, as

Additional MS, made admirable efforts to bring credibility to the MS Branch. We need more of such officers to man positions that affect the Army in the long run.

It is understood that the army has to be command oriented, as such, command assignments must have the appropriate weightage. But how much is right? How does an army commander, who is the SRO of a divisional troops' unit commanding officer do justice to writing his report? How does the Chief do justice to writing the reports of scores and scores of Major Generals, many of whom he never sees in the reporting period?

Repeated weightages for the HC and equivalent courses and the NDC, like the reservation policy, effectively kills those who for many reasons miss this nomination by a small whisker. In this system, they have no chance in hell to make up that disadvantage. Does it make sense that an officer nominated for HC Course based on battalion command at 16 to 18 years service, keeps getting hefty weightage for HC for promotion to brigadier, for nomination to NDC, for promotion from brigadier to major general and then for promotion from major general to lieutenant general? And this last obstacle clears the way for even becoming the chief of the army, if one has the age.

I have seen commanding officers who should have been chargesheeted for serious dereliction of duty, nominated for HC course, for no other reasons, than that their superior commanders were birds of the same feather. In this environment where reports are being written by different persons, the human element plays a big part. How often does it happen that there are either none or very few officers nominated for HC or the NDC course from some formation or too many from some depending on who is the senior reporting officer?

Theoretically, the General Cadre includes Brigadiers and above, practically Brigadiers' selection is based on pro-rata system as such is not part of it. This is creation of the Infantry to garner more vacancies, from a sense of insecurity. This pro-rata has spawned mediocrity at the expense of merit from other combat arms.

In particular, the bottom thirty percent of infantry officers being promoted to Brigadier, are incompetent to command brigades in combat. We are getting away with substandard Brigadiers because we have not fought a serious war for long. The low intensity conflict (LIC) in the Valley and the Line of Control (LOC) does not challenge the professional acumen of a commander, the way a conventional war will. Compare the intensity of the LIC that we deal with and the ones in Iraq, Afghanistan, Syria, Chechnya, etc. In all aspects, such as the nature, intensity, magnitude, margin of error, uncertainty, operational planning and professional intellect, time pressure, challenges and stakes for poor decision-making, leadership and personal example, and professional challenges for a commander, there is simply no comparison with conventional operations. I have seen poor commanders go up the ladder of promotion, simply because they had good battalions under them which delivered in counter-insurgency area, the so-called quantifiable results (number of kills), despite poor personal display as a commander and a leader.

A Lt General of infantry gave the justification that the infantry suffers the most risks and hardships during the career and thus needs to be rewarded and compensated, called as 'deprivation factor' in the army.

As an infantry officer who served more in the field and combat situations than most infantry officers, I have suffered more hardships and risks than most infantry officers. On that account, I could not agree more. But what is the right compensation? Should this compensation put national honour and lives of three thousand men in danger by promoting the non-deserving to the flag rank of a Brigadier? The men's lives are too precious to be sacrificed for the welfare of officers. We will pay heavily because there is no place to hide flag officers in war, be it command or staff.

Many army commanders from the mechanised infantry arm have repeatedly represented against the major shortcomings in the promotion policy that is killing merit in the name of a flawed pro-rata policy and flawed interpretation of the A.V. Singh committee report, to serve the interests of the larger arm, infantry, at the cost of the Army's effectiveness.

There was a seminar on MS Matters organised during the tenure of General Bikram Singh, the then COAS. Alarmed at the discontentment, he ordered three studies. One—Review of Quantification Policy, two—Review of Pro-rata Policy and the third of which I was the chairman—'Aspirations and Management of Non-Empanelled officers.'

None of the studies saw the light of the day. All three studies and the one by Lt Gen Parnaik, the Northern Army Commander, on promotion policy, which was believed to be a very professional and balanced effort, have been forgotten without a word. In this instance, all studies were concluded during the tenure of Gen Bikram Singh itself. The next Chief, Gen Dalbir, who was the Vice-Chief then has also retired without acting on it. Three years down the line, all that has gone into the cold storage because of narrow arm loyalties that subordinate organisational interest to narrow parochial interest. What then was the purpose of the studies ordered by General Bikram Singh, the Chief of the time? Wasting the time out in a soft kill mode?

During that seminar, Gen Bikram Singh said that Gen Parnaik's report could not be implemented because there was no consensus on it. Was there a consensus on the present promotion policy? Chiefs are there to make decisions. If everything had to be done only when a consensus is reached, we won't need the Chiefs, a computer or head clerk would do. Simply, there was an eyewash of an exercise which was doomed to be consigned to the cold storage like many others, even when it was ordered.

As the chairman of the study, I made a series of presentations. The final presentation was to Gen Bikram Singh, and the Vice-Chief, General Dalbir Singh who became the Chief later.

I said that I was detailed as the presiding officer of this study, because I had asked three uncomfortable questions to the panel in the seminar to which they had no answer. These were:

- If a Brigadier is a General Cadre rank, why is there a pro-rata on promotion to this rank?
- If there is no pro-rata envisaged in the Major General rank, why then did the AG's Branch issue out a letter giving arms-wise allocation of Major General vacancies?
- Is it fair that within the combat arms in the General Cadre, in the infantry, the officer who is at 6 on a scale of 10 is making it to the Brigadier while in the Mechanised Forces, an officer who is at 8.5 is not making it. And, above all, is it in the interest of the organisation? Or are we convinced that there will be no war?

During my briefing, I highlighted various issues. The major point that I made was that the pro-rata system was killing merit for a narrow arm-based parochial interest of the Infantry. That year, the infantry had the approval rate (PRV) of Colonel to Brigadier rank at 6.3 or so when other arms were at around 3. Over seventy officers from infantry were approved for Brigadier and many of them were surprised.

I categorically stated that undeserving officers are being promoted to the rank of Brigadiers in infantry at the expense of meritorious officers in other arms, and the sheen of this star rank has been lost. I further added that the environment does not take infantry Brigadiers seriously. He was not amused. It was my business to tell him the truth, even if unpleasant. According to my study report, a very high percentage of officers did not think high of the bottom forty percent of the infantry officers being promoted to the rank of Brigadiers. The findings of my study only endorsed the HDMC study findings on the same subject, a few years earlier, which put the percentage of officers disenchanted with the infantry Brigadiers' competence at an embarrassingly high percentage, which is rather left unsaid.

One well-read Infantry Lt General who visited my formation admitted honestly, that there was one officer from his unit, who does not even deserve to be an officer, has become a Brigadier.

The fence has started eating the grass in the army. Who will bell the cat?

On what basis are Major Generals being promoted to Lt Generals? In the given environment, only the third divisional commander is put through an exercise with troops. This means that two-thirds of the Major Generals are being promoted on powerpoint presentations made by their staff and peacetime chores that do not reflect combat ability as a commander and a leader.

I interacted with and discussed various issues with Lt. Gen K.H. Singh, a vastly experienced and conscientious officer while writing this book. His opinion on this issue is reproduced below:

> *"How do we define calibre, merit? Merit to me for Maj General will be a person who is mature, courageous, with integrity, physically and mentally fit and above all have been in operational situations with regular success. It is not about anything else.*
>
> *My experience in MS branch is that reporting was quite fair till quantification came into being. 360 degree input may be an answer."*
>
> *—Lt Gen K.H. Singh.*

In our context, this definition leaves many questions unanswered. Most other arm officers never get the opportunity to command a formation in field/ operations, these being reserved for infantry officers. Thus, merit will need to have a more inclusive definition.

Does the Indian Army have a chief or do we invariably have chiefs of their own arm or regiments? This is a comment I often hear junior officers make. How many service chiefs have left a legacy that the posterity remembers with pride? The present ACR system has made the chief of the army staff an unbridled kingmaker without accountability. The chiefs have not inspired confidence in their fairness. Those who understand the dynamics of the weightages of the three-tier reporting chain will understand what it means when the chief, as the senior reporting officer grades a Major General officer 8 or 9. In the cutthroat competition where approval or non-approval are decided on the third decimal place, grading of 8 is a known outright rejection report for a Major General. And more importantly, the chief is not required to justify why he upgrades or downgrades an officer's grading by subordinate reporting officers.

General Bikram Singh wanted to endorse/see the ACRs of all Brigadiers. What for, may I ask, when he was not in the reporting chain, nor in a position to objectively comment on a Brigadier's report because he would not know the majority of them? Whom was he trying to monitor and for what?

It is a belief in the environment, and some officers who served in the MS Branch in senior positions confirm, that the practice that was followed by some recent chiefs was that the Military Secretary (Head of the MS Branch that deals with officers promotion and posting management, etc.) would discuss the list of officers being considered for promotion with the chief, with the MS merit. The chief would decide who gets promoted and who does not. This choice is made less on merit and more on personal and regimental equations. The chief then grades the officers accordingly, 9 to those who he wants promoted and 8 to the rest. In most cases the CRs of officers in the rank of Major Generals due for promotion boards were written by the chief before the board accordingly.

Sadly, the MS, rather than advising the chief on what is ethical, colludes to help him do the unethical. No wonder, the MS is always a handpicked man of the Chief. Where is the moral courage and integrity? The MS branch declined to inform the officers seeking the date on which their CRs were initiated by Gen Bikram Singh on grounds of confidentiality. What is the confidentiality in the date of initiation of the CR? The intention is to conceal the violation of instructions by the Chief with respect to endorsing the confidential reports. This was the reply given to me by the MS branch when I sought that information through the RTI. In my case too, my three reports spanning twenty-one months were written by General Bikram Singh together within a short period. How does the Chief justify reporting on officers fairly by following this practice and there are forty odd divisional commanders. How does a Chief justify downgrading or upgrading lower reporting officers assessment when he has never seen some of them in command? All this was done to fix the merit of the promotion boards of Maj Generals, which defacto decides who becomes the army commander and the chief in due course.

The solution goes back to the point I made earlier; the present ACR system as a tool to judge the character of a flag rank officer is seriously flawed, and must be set right to get the right officers up. When the Army repeatedly quotes that less than two percent officers get promoted to the Major Generals rank and even lesser to higher ranks, should we not have chiefs who command greater respect and admiration than has been the case consistently in the last two decades?

Are promotion boards a farce?

Promotion boards to Major General and Lt General are open boards, meaning the officers are discussed by name. While 95 percent marks are allotted to various aspects such as command assignments, staff assignments, course performances, gallantry awards, etc. there is one category, called the value judgment, allotted five percent marks. This value judgement was an excellent concept to provide the officers merit for character and spoken reputation, something the ACR does not cover effectively. This has become a farce, a tool to manipulate the merit. Repeated heavy weightage for HC and NDC courses on promotion boards puts the rest out of contention.

There is a belief in the environment that the promotion board from Major General to Lt Generals is manipulated to engineer a line of succession to the chief or eliminate someone on one or other basis, despite the merit. In this category, those from the arm or the regiment of the chief or his acquaintances get the benefit. It happens that the chief says in the board, that we shall not tinker with the merit prepared by the MS branch. This takes the value judgement out of the equation, defeating the purpose for which it was initiated. The merit gets manipulated by the grading given by the chief for which he is not accountable to anyone. Far too often the chiefs have come in for criticism for unfairly pulling up their cronies. Then, while sticking to the MS branch merit sounds innocent, the match has been fixed already. It is here that the army commanders need to play their part honestly and fairly as board members. They owe it to this organisation to have the courage to do so.

Admiral Arun Prakash, the former Naval Chief, in his rebuttal to the article by Mohan Guruswamy, titled 'Flawed Generals, Who will bell the Cat?' recommends, that in cases where the chief downgrades the assessment of lower reporting officers, he should give full justification to the ratee. And, it must be given well in time for the affected officer to seek redressal, if he thinks he has been unfairly dealt with.

I wish to cite three cases that highlight the folly of our system as far as the promotion of General officers is concerned. Major General Vinod Pillay, 1979 batch and Major General J.S. Menon, 1980 batch, both from Brigade of the Guards, and Maj General Rajender Yadav, 1979 batch, Rajputana Rifles, all of whom I have known since the NDA, stand out for their unmatched professionalism, competence, moral courage, integrity and commitment to the organisation.

I would like to ask the MS Branch to show me one officer who stands anywhere close out of those promoted to Lt Generals from their batch on any count—character, competence, courage and commitment to the organisation. I am aware that these officers were not even discussed in the board. These were the biggest organisational loss. Two of these were possible candidates to be the chief and I have no doubts they would have made outstanding chiefs and made a difference to the Army. Was that the reason to sideline them?

The board results shocked the environment and further reinforced their lack of confidence in the Army promotion system. Where is the so-called value judgement and spoken reputation when such outstanding officers find no justice? I have reasons to believe that the whole procedure had lost credibility.

The army commanders are given the details of the agenda, i.e. the list of officers being considered for the promotion board well in advance. They have huge responsibility and are expected to find out about the officers under consideration, if they do not know. Yet, if the non-empanelment of officers like Maj General J.S. Menon and Maj General Vinod Pillay, are not even discussed in the board, then the system has become irrelevant. These are not exceptions, it happens repeatedly.

Unless you have truly impartial chiefs of great morals and conscience, they do not deserve the privilege to have unaccounted authority to determine the fate of General officers. No wonder there is always a spurt in promotions of officers from the chief's regiment or his clan.

What must be done?

> *"Conformist mediocrity will always win over non-conformist brilliance in the Indian Army"*
>
> *—Brig Sanjay Aggarwal (Retd)*

This is a sad commentary on the Indian Army by a well-read senior officer.

Sadly, the ACR in its present form has become redundant. Like the straight trees, those who have the moral courage and integrity are most likely to be cut, and the system provides for many soft kill methods, where the victim never gets to know what has hit him. It is my belief that barring those who get caught in acts of corruption, every Brigadier and above are graded 9 in integrity and moral courage. Does that reflect the truth of standards of moral courage and integrity in the higher ranks? No, it does not.

Two divisional commanders commanding their divisions on the border in J&K called me up for petrol when the Army went through a fuel crisis in 2013. My division had elements both in Meerut and New Delhi. I was surprised, asking me for fuel in New Delhi when they were in J&K? They wanted petrol for the Army vehicles in New Delhi left for their wives, in violation of all ethical norms. I politely conveyed to them, I can't be party to their crime. Both these officers were decorated and promoted to the next rank.

To me it is not trivial; it is a matter of lack of character and integrity. The line between perks and corruption is very thin. Sadly, senior officers draw their own line as it suits them personally. Do such officers qualify to judge others on integrity? Such is the truth of our promotion system. A dishonest commander is most unlikely to promote an honest subordinate. Such commanders render themselves weak and vulnerable, thus are forced to look the other way when a subordinate is involved in acts of misappropriation and corruption. The value judgement was meant to redress these anomalies. Their seniors were either ignorant of what these officers were doing or were incapable of acting against them for similar reasons under the false garb of perks and privileges. Will 360 degree input help? I am sure it will.

Can one believe that two Chiefs, the Army and the Naval Chief of the time and the area commanders would not know that the Adarsh building, just on the entry gate of Colaba military station, was not on military land, and now the former Air Chief is in the news. Simply, the deal was too lucrative for them to say no and they chose to subdue the voice of their conscience and let the Army down. We must remember:

> *'The voice of conscience is so subtle that it is easy to ignore it.*
> *Yet, it is so loud that it is impossible to mistake it.'*

We always know when we are being dishonest, illegal, unethical or unfair.

Interestingly, the patterns of grading officers in the CR across ranks are monotonously identical. All star qualities are graded 9. The columns on physical attributes, communication skills, knowledge of other arms and services and administrative abilities are graded 8 and juggled in a different manner.

If the Army wants men of character and integrity to the top appointments, it will have to muster the courage to introduce a 360-degree assessment on character and integrity. This idea has been discussed earlier, but the Army has found a fig leaf to hide and prevent its implementation. The fear of populism is ill-founded. The junior officers are a highly dedicated and proud lot. Barring insignificant exceptions, they will not vote for morally corrupt or dishonest commanders, no matter how populist they were.

I was the proponent of this philosophy even while in service.

My recommendations

The studies conducted during the tenure of General Bikram Singh in 2014–15, which by design or default went into the cold storage, must be resurrected and implemented in the best interest of the organisation, fairly and in a transparent manner. There are practical and fair suggestions on pro-rata, weightages for courses and other important issues.

The policy must ensure that only officers of strong character and integrity are allowed to rise, for they set the standards for the posterity.

For all boards for promotion to Maj Generals and Lt Generals, there should be a peer ranking by the agenda. For example, if there are forty officers in the board, they should be required to group the officers, less themselves, in brackets of five from top, in their assessment. The fear that regimental and other affiliations will come into play are ill-founded, as the merit will be thrown up when being graded by such a large and diverse group. In fact, it will expose those who resort to cartelisation. This input must be counted in value judgement alongwith the 360-degree input.

I am convinced, that the present system does not reflect the true character and integrity of senior officers, causing embarrassment to the organisation. To get this right, the Army must institute a system of 360-degree input on the character traits of the senior officers.

The methodology can be worked out. One suggested method is that when a Major General comes up for his board, which would be about two years after he has relinquished his command of the division, confidential inputs from one or two of his brigade commanders and three/four colonels of the division, that could include officers from staff and commanding officers of that time be obtained based on a questionnaire on his character traits such as integrity, probity and leadership, etc. as is being done for the selection of defence attachés. These should be the tools for value judgement; A similar methodology for Brigadiers who are likely to make the grade to separate the men of doubtful integrity.

I would like to mention here that when an officer is selected as defence/ military attaché, the ACR is only a tool to shortlist the officers. Their selection is not based on ACRs alone but on additional inputs that the MS Branch seeks from the same superior officers who have initiated their reports recently. These inputs are about those important character traits that are not covered adequately in the CR format.

While the organisation feels the need for additional inputs for selecting colonels, it does not do any such thing even when selecting officers for the rank of Lt Generals. Who is more critical to the Army, a DA in the rank of a Colonel or a Lt General who could be the chief one day? The Army must have the courage of implementing measures that expose corrupt and self-serving officers. There is far too much at stake. The war comes once in ages, but when that happens, we cannot afford to fail or get the men killed because of poor leadership—which is all about character and competence.

In the period 2013 to 2016 there were repeated successful attacks on the military camps. How and why are the terrorists succeeding repeatedly when the threat is clearly written on the wall? Where is the culpability of the commanders whose business it is to see around the corner and anticipate the unexpected, be a step ahead of the enemy and ensure that the battalions are prepared for the challenges? While the challenges are well appreciated, it is precisely this reason that dictates that there can be no charity and only the most deserving must be promoted. Again, I ask the same question, are we professional as we claim to be?

And to highlight the point above, as the sector commander in the challenging Lolab valley in 2008–2009, my sector killed the largest number of terrorists in Northern Command. Not one man from my four battalions got a scratch in operations in eighteen months. Get the right man up and you will save precious lives and embarrassment.

In the neighbouring sector, the commander asked me to give a motivational talk to his commanding officers. I was rather surprised and out of my wits at this strange request. I cannot imagine myself asking someone else to give a motivational talk to my commanding officers. It is admission of his own incompetence as a leader. During an interaction with him, I asked him how much was he involved in ensuring a result-oriented training in his sector? He said, he had left it to the commanding officers. During the army commander's visit, he was told by the army commander that he had not come to listen to the JC pamphlet points. He wants to know how he is making a difference as a commander. No wonder, in one operation in his sector, one unit lost nine men against eleven terrorists killed in a day. But because he had one good battalion that carried him over their shoulders, he lived to serve in the next rank. The Army must hold commanders accountable for professional inadequacies.

The challenge today is to select senior officers in peacetime environment who will deliver in war. The present system of promotion lacks even part assurance that it does. Don't we see the officers seen as good command materials get promoted for staff and those who inspire no confidence getting approved for command.

The starting point is to set the allotment of vacancies right at the rank of Colonels. The Rashtriya Rifles and the Assam Rifles battalion must be all combat arm vacancies. The rest will take care. It is also important in my view, that, given the size of the infantry and its commitments, there has to be a reasonable representations of infantry officers at the level of corps and army commanders.

The Indian Army is shy of taking action against senior officers both for incompetence and moral turpitude.

Thomas Ricks states, "We often think of the military with a culture of clear accountability. This is only really true for lower ranks."

Case Study 1

In 2001–02, the commander of an independent armoured brigade was chargesheeted for altering the ACR points of the commanding officers of his brigade, after showing the report and having obtained their signatures. Once the matter came to light, the MS Branch asked the officers to confirm whether this was the original report. This commander tried all ways, including through his spouse, to get the officers to bail him out by saying that this was the original report.

An act of grave misdemeanour in every aspect by an officer of this rank, he should have not only been sealed for good, he should have been handed an exemplary punishment. Ironically, the punishment awarded was of little significance and he still hoped to become a General, and interestingly, was disappointed when he did not become one.

The reason, as I understand, was a superior deeply obliged to him. How did this man become a Brigadier in the first place?

Case Study 2

I have quoted two well decorated Generals who were in the news from the Army Service Corps and the Ordnance Corps. Approved for Lt General, apparently in line for a PVSM, placed under investigation for corruption. How did they reach this rank?

Case Study 3

Major Generals Vinod Pillay, J.S. Menon and Rajender Yadav are outstanding command material, high on moral courage and organisationally oriented. Everyone who knows them swears by them for their character, competence and integrity. Why did these outstanding officers not make it to the next rank?

The three examples above clearly highlight the grave shortcomings of our system. Who will bell the cat?

The MOD wide their letter dated 17 Jan 2017 constituted a committee to review the Promotion Policy for the Indian Army. It had a vide scope that included major important issues. It asked for recommendations to review the allocation of vacancies and Quantified System of selection to infuse more objectivity, transparency and fairness in the system.

I have the following observations:

- MOD interference is a result of the failure of the Army headquarters to act on repeated representations from within the Army, resulting in recourse to the courts by officers enmasse. Unless we put our house in order, we may lose the executive control of the promotions of star rank officers to the MOD. And that will be the last nail in the coffin of the institution of the Army.
- The directive, while mentioning the allocation of vacancies, has by default or design omitted the specific and important aspect relating to pro-rata concept in General Cadre which had no government sanction. So, who drafted this letter—certainly not the MOD. It must be the MS Branch or the AG's Branch and they have omitted an important issue.

- I trust the credentials of the two Generals to do a fair job. However, they belong to infantry and artillery, the two beneficiary arms of the current promotion policy. The Mechanised Forces, which have the most severe grievances for justifiable reasons, have yet again been left out. Their representations have been ignored in the past. I do hope, their recommendations are suitably taken note of now.

Professional Army: A Reality Check

Why did Kargil happen? Why did we have to learn from a shepherd of the intrusion? Which senior formation commander was held accountable?

It is my unshakeable belief that the Indian Army will always rise to the challenges that may be thrown in its line of duty. They always have, they always will. This belief is born out of the strength that our units provide as the rock solid basic building blocks of our Army. They will because of the regimental ethos, the sense of devotion of our young officers and their readiness to take risks and make the ultimate sacrifice. They will because our men are excellent soldiers when well led. They will because our young officers are excited by the thought of combat and their best comes out there. They will because they know no other way.

But true professionalism lies in prevention rather than cure. It lies in accomplishing the mission in the most cost-effective manner in terms of time, resources and human lives, when forced into it. I believe, this is an area that is a weakness.

> *No bastard has ever won the war dying for his country,*
> *You win the war making the other bastard die for his.*
>
> —*General Patton, US Army*

At the highest level, the acme of leadership lies in winning without fighting.

Profession means a vocation or a job that one undertakes. 'We are a professional army'; we have heard that a thousand times. When using this term, we are trying to convey a sense of expertise and mastery in our vocation, and our belief that we are great at what we do. What does this really mean?

A professional is a man who is an expert, specialist and proficient at his job. A professional sportsman is one who makes his living out of sports. To be a successful professional sportsman, he has to be amongst the best in the world. To survive and earn his three meals a day, he has to constantly hone his skills, strive every single day. He has got to dig a well every day to get a drink.

In the context of our profession, my perspective on professionalism has been stated clearly earlier. Professionalism has as much to do with character, passion, emotions and selflessness as much as with competence.

Where are we on this today?

I wish to quote an excerpt from the book, *War and Soldiering*:

> *"The system is progressively losing to absorb dissent, withstand honest criticism, tolerate uprightness even if inconvenient, and to promote greater camaraderie and a sense of moral obligation. It is both unable and unwilling to adapt to warlike conditions and their particular requirements, leave aside respond spontaneously and positively to its topical contemporary challenges in the vital sphere of man management. It is increasingly confirming itself as a peacetime army, with peacetime procedures and ethos, predominating even in warlike environment under the cussed cliché—'time tested'. At this rate the army will not fight the next war as it should, notwithstanding the "glorious traditions" and the smug façade of the holy cow dished out by the complacent government and lapped up by the gullible public."*
>
> *Source: "What Ails the Indian Army Officer Class?" Lt General S.C. Sardeshpande,* published in www.indiandefence review.com dated 21 Nov 2015.

The Kargil war saw a disproportionate number of young officer casualties in comparison to JCOs and men. When the number of JCOs present in the unit is almost four times the officers, the statistics throw up startling revelations that must be taken note of and addressed. We are proud of our junior officers who faced the brunt of the battle. Without their 'devil-may-care' attitude, it would have taken much longer to win this war and at a much higher cost. But then, there is a serious lesson here—that our young officers possibly did what was the job of the JCOs and section commanders and that is not professionalism in the true sense. A battalion is professional when all its elements are capable of performing their mandated tasks and those of one level up, in a fluid situation of uncertainty. It is professional when it can function under all seconds-in-command.

The general impression in the Army is that the JCOs' performance leaves much to be desired. Compare this with the deliverance by sergeants in the US and European armies. Clearly this falls short on professionalism. The JCOs must be trained well, given full authority and responsibility and held accountable. It must be understood that the bulk of the functioning of the unit at JCOs' level is repetitive. He has been doing that for decades. There is simply no excuse for not being able to perform the routine tasks satisfactorily. It is the professional business of officers to ensure that they deliver and earn their pay cheque.

As the sector commander in the Valley, I would occasionally debrief the JCOs returning from operations. I was shocked that one artillery JCO had no idea of map reading. He could not show me his company operating base (COB) on a metremap, that had been in the same location for years, and he had been there for over a year. The commanding officer tried to justify that he knows the battalion area backwards and does not need a map, and that he was a driver by trade. Has he not passed various tests to become a JCO? Will he never be required to operate outside his unit area of responsibility? Is he not expected to know the basics of map reading and navigation if he has been placed in command of a platoon? For God's sake, give me a break.

Our approach to training inspires little confidence. The manner of training of an average unit is pedestrian. Passion, purpose and application is usually missing, resulting in going through the motions.

I find it embarrassing to mention the statistics, but the state of physical fitness of young officers is a cause for concern. The obesity rate of middle rung officers reporting for courses at the Infantry School is disheartening. The failure rate of potential commanding officers doing the senior command course in physical tests does not inspire confidence. This factor relates to the basic requirement of officership in the Army at the cutting edge—physical fitness, which should be a given and not a cause for concern. And it is disheartening.

One divisional commander was impressed by the standards of physical fitness and firing of the units reflected in their annual training reports. He had seen enough in his career to see through the inflated figures.

He ordered a board of officers to conduct the physical and firing tests of the division in accordance with the Army orders. Results; a mind-boggling percentage of the division failed in physical fitness and even more failed in firing. Another division commander told me the same thing in 2017, but specifically withheld the figures for fear of being quoted. Yet another facet of professionalism, a commander afraid of being quoted of the weaknesses in his division! I bet, the results of the most divisions would not be much different. Yet, most commanding officers, nearly all brigade and divisional commanders, keep getting graded outstanding.

As a divisional commander, I asked my Flag Staff House Guard to 'stand to.' They were taken by surprise and had no idea what to do. Clearly, the unit lacked the sense of what they were there for, even though all guards in my division were tactical and my operational orientation well known to my division. One soldier got so entangled in the rifle sling that a terrorist would have slapped him and walked away. And these were the selected and trained lot. I am sure this is not an exception.

In 2002 in Jammu, terrorists infiltrated into an infantry battalion, killed the soldiers in the quarter guard, *the sanctum sanctorum of security and alertness* and escaped unscathed. The Indian Army be praised, the commanding officer was nominated for HDMC. Professionalism? Try other places, and I bet my last ten bucks, the majority will be the same.

Every guard is there for a purpose. It is there to protect something. Protecting an asset is not a function of being rooted in the sentry post, it is a function of controlling the area of interest.

There is far more attention to be paid to basics. It must become a habit of every detachment commander to analyse his task, what it involves and rehearse it.

The year 2016 and 17 saw a spate of terrorist attacks on military camps. Despite a high level of existential terrorist threat, and despite a bulk of the troops being on security duties, we were surprised every time and suffered heavy casualties. How is it that the enemy can repeatedly behead our men and get away? How are these teams operating that it happens repeatedly? Before the dust is settled, citations are written, awards are announced and the basic shortcomings are forgotten. The cycle goes on. Bullshit has replaced substance. It must be curbed ruthlessly.

If the Army is shy of issuing ammunition to the guards and protection teams, there is a serious problem in our functioning. I know of a Rashtriya Rifles Force Commander, in whose QRT ammunition was kept with the QRT commander and not issued to the soldiers for fear of untoward incidents that could affect his report. Where are we headed? However unpleasant it may sound, we are far from where we should be.

Powerpoint presentations will not help. All combat happens on ground and not on sand models. And sand models cannot factor the realities of war—leadership, terrain, competence and enemy.

What does it speak of the professionalism of our units? What does it speak of the vision of our officers whose business it is to be able to see the invisible, be ahead of the terrorist in the mind games and be prepared for him? What does it speak of our senior commanders whose business it is to ensure that their units are on the ball?

We have done enough exercises in various capacities and seen the Army across all spectrums in operations and peace. In the present state of affairs you will continue to pay a far higher price to achieve your missions. Unless the basics are thorough at the sub-unit levels and the senior ranks have the intellectual depth and the vision to see around the corners, success will come at a heavy cost. It will only get worse as the nature of warfare and technology works to the terrorists' advantage. I am afraid, in my assessment, our basics leave far too much to be desired.

Our commitment to training lacks passion and application at all levels. Quite often it amounts to eyewash. One of my commanders and his commanding officer conducted me through their training area. I was aghast at the soldier attempting to bayonet the enemy through the loop hole on a DCB bunker. I can quote hundreds of examples.

My posting order as a divisional commander came five days before date of moving out. It meant that my relief was arriving the next day. The same day, I visited a unit on night training. After I returned, I gave a call to a Lt General, visiting his unit in my formation, who later became an army commander. He was surprised that I was out on night training with troops after my posting order was out. And he was a good officer by all counts. My point: I was out because my troops were out in the cold through the night. I was out because I needed to be satisfied that they were training in the right direction which would save lives somewhere. My visit to the night training cannot have a link to my posting and ACR. It was about conscience. It was about professional integrity, doing the right thing as the commander, even if no one was watching.

Unless the commanders have this kind of passion, commitment and ownership to their responsibilities and their men, professionalism will be a far cry.

Where does the senior hierarchy stand on this?

The senior hierarchy is dogged by careerism and narrow focus aimed at getting two outstanding reports in their tenure. That malaise starts from the commanding officers and is worst Brigadier onwards.

All that is needed is a study of how much application of mind is taking place at the divisional and corps levels in formulating operational plans. We are promoting our senior commanders on powerpoint presentations and peacetime-based assessment. Is this methodology enabling selection of commanders who will deliver in battle? The last two decades have seen immense change in technology, capabilities, terrain and many factors that impact operations. But we are still coasting along as we did ten years ago. I was on staff in a strike corps as a colonel. I landed in the same corps as a divisional commander seven years later. To me it seemed, nothing had changed in the thought process and its application, despite changes in weapons, terrain, threats and technology.

Professionalism flows from professional integrity. Where are we today?

Our senior commanders, barring a small minority, suffer from a rigidity of ideas, absence of innovativeness and lack of mental flexibility, intellectual draught and risk-averse traits. Unfortunately, war is all about uncertainties, seeing the invisible and the ability to respond to uncertainties through pro-active bold actions. From what I have seen over decades, majority just do more of the same—walk the beaten path: monkey see, monkey do. Their comfort zone lies at the tactical level, in status quo and following the established pattern; all serious drawbacks as a leader in the glorious uncertainties of war.

This throws up many important issues.

In the above case of false reporting of professional standards, were the brigade commanders doing their jobs honestly? If they were, the state of affairs should have been better or they should have taken the unit commanders to task.

Are they party to fudging reports? If they have allowed that report to be forwarded to the divisional headquarter, they are a party and fully culpable.

What about the commanding officers? They are at fault on two counts—one, failing to train their command to the required standards, hence incompetent. Two—fudging training reports. Both serious offences as commanding officers.

As far as the senior hierarchy is concerned, one only needs to honestly look at the way our operational discussions and exercises are conducted. Little good comes out of all this because of the manner in which they are conducted. No enemy ever captures anything worthwhile in an operational discussion of a holding division, while own offensive reaches the heartland of the enemy, bringing out wrong lessons and defeating the very purpose of these discussions.

Our mobilisation timings of mammoth formations like the armoured division, be it winter or summer, are telescoped to tailor fit the need of the discussion at the cost of realism. Preparatory timings in the assembly areas have been reduced to what would suit a computer game.

What did I do in my time, would be the next logical question.

I pointed this out to the corps commanders as the Col GS of a corps, and later as the divisional commander in the same corps.

For many reasons, some out of the Army's control, these exercises have little training value. They become stage management exercises to impress the VIPs where all professional integrity is thrown out of the window. The lack of training areas and other constraints mean that if the Army was to go to war, the formations will be in for surprise.

A division with its operational role in developed and obstacle-ridden terrain, doing its once in three years training in the desert and marking canals and rivers by tapes is an entirely wasteful effort that does more harm than good. One only needs to see the reports generated after such exercises to see the eyewash.

I was the chief umpire for a strike corps division exercise. What I saw and pointed out to the corps commander was not encouraging at all. The division commander was more concerned that his commanders get a good dinner rather than be with their formations assaulting across a river. With no uncertainties and elements of war, this division was way behind schedule everywhere. The armoured brigade was waiting for induction in a drill square layout at midday, a perfect target for destruction in one strike. Stories of other exercises are no different.

Try these simple tests

Give an engineer regiment a task to construct a bridge *strictly under operational conditions* of reconnaissance, management of stores, time constraints and simulated enemy interference through air, artillery and long range ground-based weapons as can be realistically expected in a modern-day battlefield.

A similar exercise for an armoured brigade to induct over the bridgehead by night in operational conditions. You will know where we stand.

A mobilisation exercise for an armoured brigade in summers and winters. You will soon know how we are fooling ourselves in sand model-based operational discussion.

Commanders will be overwhelmed by the quantity of equipment on the ground they have never had the opportunity to handle in peacetime training. In the high technology battlefield with a competent adversary, a disaster is beckoning, unless we get our act together.

Our assessment system does not identify suitable war time commanders—who will succeed in uncertainties in war. It is a predictable pattern which has become stereotyped and needs to be corrected.

Missing the woods for the trees

While we are talking of cold start and pro-active strategy, the Indian Army at all levels is becoming increasingly bulky and heavy and thus less agile. The headquarters look like small cities. One of the strike corps in which I was Col (GS) in 2006 had a mobile operations room of three modified vehicles that could be set up in a reasonable time. Three corps commanders later, it became an operations room with nine modified vehicles, with an exotic name, that took three days to set up with immense logistics to support it, that cannot be concealed in today's battlefield. When we talk of short wars, where is the place for an operations room that takes three days to set up, and that too for a strike corps formation, which is expected to be lean, fleet-footed and mobile. The whole concept kicked the secrecy of information of an offensive formation out of the window, with every Tom, Dick and Harry getting access to the activities they had no business to be witnessing. Sadly, that is the culture today; do something for doing sake and leave a legacy by creating five star structures.

When I questioned this, I was told that this was only for peacetime exercises. Great! You are practicing one thing in peace and will do something else in war. And not practicing what you will be doing in war. Some logic our people have!

One corps commander complained to me that when he stayed in one of my brigade guest rooms in Delhi, it was not up to the mark and he was not looked after as a corps commander should be. I said, apologies for the human element, but I am a strike corps formation. My KRA is operational readiness and not five-star guest rooms. Four months after I left the formation, my former corps commander called up to tell me the same thing, asking why did I not tell him this? I told him this was no crisis. I was a division commander; I do not come running to you for everything. I wondered, four months after I had left the division, the two corps commanders were only discussing the state of guest rooms and how they should be looked after. This is how officers seem to be appraised these days—event management and peacetime chores.

I attended an exercise with a Russian army commander in Russia. It was just him, me and a driver. No second vehicle. No organised food or water throughout the day. Carry your own. When we returned after the day out, one soldier handed him a plastic bottled water. That is it. This is part of a professional culture. Much against the popular perception, the Russian army is very strict when on training. No vodka – no liquor during training, however long the duration.

I served a ten months' tenure as the DDG, NCC Delhi Directorate in 2010–2011. The first thing I did was to study the NCC syllabus. To my horror, the syllabus that had been formulated post-independence had never been revised. We would have had three dozen DGs NCC and ADGs NCC in the NCC Directorate headquarter by then. Each state was headed by a Brigadier, later upgraded to Maj General, and hundreds of officers as group and battalion commanders. The right syllabus is the basic thing on which the output of such an institution is dependent on. It had lost much of its relevance, but was carrying on. It never occurred to anyone that it must be revised, simply because it needed effort and commitment. Everyone was happy poodle faking. I initiated the syllabus review, which was eventually introduced. That is just another example of lack of professionalism and commitment. Many DGs had earned themselves decorations in this appointment doing little of substance but of immense publicity.

In the armoured squadron (NCC) of my directorate, the armoured fighting vehicle, meant to be used for cadets' training, was buried in the ground till the turret. So much for our professionalism.

Years later, after I left the NCC, the ADG of my time introduced me to one of my old commanding officers as the man who changed the concept of NCC in India.

I cannot believe that it occurred to no one in six decades that the syllabus is out of sync with the current environmental needs and needed a review. Whether it did or not, either way, it is a sorry story on the professionalism, integrity and commitment of officers who served in the NCC, in any capacity.

Let us introspect and answer a few soul-searching questions.

- Left to ourselves, will we be honest in taking promotion exams?
- Are we honest in our reporting or do we fudge the unpleasant?
- How honest are we in conducting exercises—unethically undertaking reconnaissance and rehearsals in actual sites that would not be feasible in operations or marking routes in enemy territory, etc.?

While some reconnaissance and stage management is permitted for public safety, it is exploited to do the unethical, just to impress your superiors, drawing wrong lessons that could bring dishonour to the Army and a loss of lives in war.

> *"You can never trust a commander who is not guided by the moral compass of his conscience."*
>
> —*Author*

Lessons

The Army has rarely been embarrassed by the junior or middle rung officers. In fact, they are the jewels in our crown. It has been repeatedly embarrassed by senior officers, including the chiefs. Far too many have sacrificed their lives for the Army and thousands others who have suffered to uphold the values of the system. Let us prevent the Army any more embarrassment. Before the MOD takes control of the Indian Army promotions, let us fix the system, rising over petty gains that only reflect small minds. Solutions to all military problems lie in having men of character and integrity at the helm, based on merit alone, men who are not guided by their personal interest but by their moral compass.

The Senior Hierarchy: Flawed Generals: Who will bell the Cat

In the last few years many things have happened that anger the serving and the retired military fraternity. The repeated downgradation of the military since independence and the indignity heaped from the third Central Pay Commission to the seventh, the never ending one rank one pension, the lingering anomalies of the pay commissions for decades, surprise discontinuation of rations, the bureaucratic manipulation of things that affect the Army and the promotion policy discontentment, etc., have been a cause for military disenchantment with the senior hierarchy of the Army. The latest nail was the order to clean the high-altitude areas of the Leh-Ladakh of the muck left by the tourists. A number of articles appeared in the media and the social media, including by a former Vice Chief. I was talking to the Deputy Chief about it, who said, it was a general order to all the ministries. A general order cannot be applied to the Army because of its implications that must be understood by the senior military commanders who issue orders. It was so disconcerting to see on 'WhatsApp' a photograph of a commanding officer in uniform sweeping the road in full public view. Do our senior officers have no clairvoyance or a sense of pride to see what a soldier should do and what he should not do? When I pointed this out to a very senior officer in the Army headquarters, he said, why should he do anything about it, the army commander in whose command it has happened should do it. And we talk of soldierly pride?

The government and the Army must never forget that the 4 Infantry Division that was routed in the Kameng sector in Arunachal Pradesh in the 1962 war was constructing barracks and growing crops under a pliant and ambitious Maj General B.M. Kaul as the divisional commander in Ambala.

Later, as the QMG, he was more interested that the troops construct shelters rather than prepare for the looming war. Our business is to win where we fight and anything that interferes and weakens the soldierly pride and ethos must be dealt with summarily.

And now there are rumours of the junior officer ranks being examined for downgradation to Class B officers. The anger of the serving and retired officers on this is alarming. If that is true, it only indicates what a joke we have allowed the Army to become. This is the last nail in the coffin of what once was a proud Army. An Army without pride is no army at all and that is what differentiates the military from the rest of the uniformed services. The blame must be put squarely on the shoulders of the senior hierarchy of the Army—the Chiefs, the army commanders and the PSOs (Lt Generals)—in that order. There are orders that need to be questioned, challenged and then if the sense does not prevail, simply disregarded. It is pretty clear where this has emanated from, not from the politicians, but from the bureaucracy.

In an acerbic critique titled, "Flawed Generals; Who Will Bell the Cat" (*Asian Age,* June 30, 2016) Mohan Guruswamy, an eminent economist, flayed the upper echelons of India's military as "visibly dense" and "obviously incompetent". These "uninspiring leaders", according to him, rose in service, by managing to "get good ACRs year after year with bland obsequiousness".

The article generated interesting debate under the "Victory India Campaign" headed by Col Dalvi, who as a dedicated officer is putting immense effort to highlight the shortcomings that need to be addressed in all spheres of the services. Large numbers disagreed with him, mostly because it came from a civilian and why he had chosen to single out the military when all other institutions from the polity, parliament, bureaucracy, the police and even the judiciary are all afflicted by corruption, inefficiency and absence of values.

There were quite a few who agreed with him. I too do to a large extent.

Guruswamy made the following points in his article:

- "Outstanding officers with a strong individuality and intellectual curiosity get culled by the stubborn seniority system.
- He refers to Norman Dixon's seminal *On the Psychology of Military Incompetence* where he poses the questions: "How, if they are so lacking in intelligence, do people become senior military commanders? And what is it about military organisations that they should attract, promote and ultimately tolerate those whose performance at the highest levels brings opprobrium on the organisations they represent?"

- The Indian military, like many others, doesn't appreciate standout talent and personality, and prefers a uniform greyness. The system beats out the commander and dashing leader in an officer long before he becomes a General. We will never study this, as if this opaque system of evaluation is a military heirloom.
- Fortunately, we have not had a major war in recent times to test the mettle of our commanders. But even in peacetime, many have, unfortunately, managed by their acts of omission and commission to bring opprobrium on our military.

He quotes Thomas Ricks, "We often think of the military with a culture of clear accountability. This is only really true for lower ranks. In contrast, there is absolutely no question that if the British Army were a listed company (heaven forbid), a slew of Generals would have been kicked out of theatre early."

As wars become shorter, we can ill-afford to test our Generals in war. We have to do it well ahead and be ready for war.

The nation cannot afford the military to be a bureaucracy where even the undeserving rise. The solution to this can come only from within the military.

The politician is not interested. The bureaucrats will just bring more of their ideas, which could be fatal. The military must look within. But who will bell the cat?"

An honest introspection will tell you that there is more than an iota of truth in what Guruswamy says. Few Generals will agree with him, a majority of others will.

Admiral Arun Prakash, former Naval Chief, in a balanced rebuttal made the following points:

Progressive inflation in ACR marking, has over the years, undermined the reliability of the system.

- In the computer generated "merit lists" based on the ACR numerical gradings, averaged out to the third or fourth decimal point, the top scorers in this "lottery" of rankings are recommended for promotion, and it should not come as a surprise, if a number of undeserving "bad eggs" slip into higher (and highest) ranks.
- Guruswamy makes a valid point about the necessity of "obsequiousness" for advancement. But this is a cultural trait in our society and we all know that age, rank and financial status demand much more deference in India than anywhere else in the world.

- Many fine officers do get eliminated at an early stage for their non-conformity or forthright views. There is no doubt that the evil of sycophancy—so prevalent in India's politics and society—will undermine the roots of our military unless the senior leadership curbs it ruthlessly.
- Another area of disgruntlement arises from the fact that the service chiefs are entitled to have the last say as far as the ACRs of Maj Gen, Lt Gen and (except in the Army) C-in-C rank officers are concerned. This is quite appropriate, and the chief's numerical grading can override any other assessment in the ACR.
- However, like other reporting officers, the chief, too, must provide full justification, via written remarks, for overruling the earlier grading(s). Given the fact that the chief has the power to change relative ACR rankings at the highest levels, there have been allegations, of late, regarding chiefs contriving or fixing "lines of succession".
- The fact that such a contention, however implausible, keeps cropping up should be a cause for reflection.
- The most damning criticism relates to the suggestion that chiefs and senior officers refrain from taking a stand with the MoD on critical service-related issues, because of the lure of post-retirement appointments.

Personally, I feel that it is an unfair aspersion, unworthy of our comrades-in-arms. Be that as it may, in order to kill this kind of unhealthy speculation, amongst the rank and file, it may be worthwhile for the current chiefs of staff committee to consider suomoto renunciation of post-retirement government posts as far as the service chiefs are concerned.

Admiral Arun Prakash disagrees on Guruswamy's point that the Army Generals are not tested in operations, stating that the Army has continuously been engaged in counter-insurgency operations and doing an outstanding job.

I have serious disagreement with the Admiral on this point. While what he says is true, we should not ignore the fact that these counter-insurgency operations, while having their own challenges, are rather low-key operations as compared to what US and coalition forces experienced in Iraq, Afghanistan, Somalia and other places. In India, the dynamics of these operations have very little in common with conventional operations. These operations test the junior leaders up to battalion commanders only, and very little beyond that, if at all. In the Valley the senior leadership is also comfortable operating at the tactical level with no long term perspective, which is their prime job. The division and corps commanders are also counting the kills as their measure of success and fighting the AK 47, whereas they should be fighting the ideology. There is clear void in the intellectual application and understanding of the long term perspective of the problem. This, in my perspective, is one of the causes of lack of sustainable progress in the Valley and the cycle of instability keeps repeating.

Some other views expressed in that debate on the article by Mohan Guruswamy:

- If Rommel, Patton, Montgomery, Mc Arthur, Guderian, Schwarzkopf were to serve in the Indian Army with its ACR system, none of them would have gone beyond the rank of Lt Col.
- An officer is supposed to be bold, aggressive and a killer in his attitude. Where does the ACR seek these battle-winning traits of an officer? Officers with these qualities, needed to be good operational leaders, have no future in our ACR-oriented army.
- You need Generals with courage to stand up and speak their mind to the political leadership. If we had such officers, we would not have given up such a strategic area as Haji Pir in 1965 or let go of 94,000 prisoners without dividends post the 1971 war.
- That the Generalship is at the lowest ebb today. The decline started after the 1971 war when competent Generals were sidelined. It kept getting worse and reached its lowest ebb when the line of succession began to be engineered.
- Command in operational areas is a ticket to professional growth; there is no risk and little challenge. See the list of awards for seemingly no accomplishment, except to have served there.
- Within the Army, the corps and regimental loyalties ensure that mediocre subordinates are pushed up.
- If one of your chiefs becomes a low med category for impaired hearing, just a fortnight before he retires, to claim 20% disability, what moralities do you expect from subordinates? He was an artillery officer. Sure, his ears were not damaged in the AC Office of the COAS. The same chief figured in the Adarsh Housing case too.
- A look at our chiefs since General J.J. Singh tells the story of the quality of our Generalship. The last two Chiefs have left a legacy hardly worth remembering.

Shobha De in her article on Lt General Bipin Rawat as the next COAS, carried on www.sanjhamorcha.com/Jan 2017, wrote this on our chiefs:

> *"Seen only hovering around and pulling chairs for his immediate bosses, the combined tenure of Bikram and Suhag was looked at by most army officers as the dark age which would finally come to an end in December 2016 with the retirement of Dalbir Suhag. The biggest casualty during this period had been the self-respect, for unless an Army is led by Generals, it is no army at all."*

Kanti Bajpai, a regular columnist, is not the man to shoot off the hips. In his article titled "Dangerous Nonsense", published in the *Times of India*, 17 June 2017, he made the following comments:

> *"No institution including the constitution and the president is above criticism in a democracy. Once we put the Indian Army on a pedestal, we will become Pakistan—and look where the unhappy country is.*
>
> *Anyone who is familiar with the Indian military knows that for all its strength, it is marked by professional incompetence, bitter rivalries at the top (as we have seen in one leadership succession after another), a feudal culture of officers and batmen (Indian Army has one of the worst teeth-to-tail ratio in the world), and, yes corruption. We pretend that it is a saintly organisation of selfless heroes, but the military itself knows that it is not so. It is a part of the wider political and social culture of India, and it is not terribly exceptional.*
>
> *It does a difficult job, which is not easy. It must get its fair share of credit and its share of criticism."*

It is not my purpose to comment on his views. Articles by civilians could lack the necessary research and in-depth understanding of the complexities and dynamics of military functioning. The Indian Army has never failed the country. The reasons for the 1962 Indo-China War debacle are too well known to warrant a repetition. But, they are not entirely wrong in all they say. We need to introspect, why is the impression of incompetence in the Army so common in the civil academia, take the bull by the horn and fix the damn problem. If it means putting own house in order, which I think is the need, must be done honestly, for our own good.

It would be a good idea to understand the perspective of the academia who have questioned the Indian Army's competence. We must understand the basis of their impression and criticism. If they measure the military competence by academic intellectualism of their own domain, it is misplaced and needs to be corrected. This notwithstanding, a senior military officer today has to be intellectually equipped with an awareness of multiple environmental, political, diplomatic and social issues, and their influence or impact on military decisions and operations, and conversely, the impact of military decisions and operations on environmental aspects.

His competency at the higher level to deliver as a military leader cannot be complete without this understanding. If they suspect the incompetence lies there, there are serious issues that the Army needs to introspect on. The Army will only get better by positively responding to their perspectives and criticisms.

I know officers will say, they know little about the Army and understand our ways even less. How are they qualified to make a judgement? Don't we have opinions on the bureaucracy and the politicians? And don't we think we are right in our assessment? The point is why and where have we reached that such impressions prevail in the civil academia.

While we may have a problem with civilians' criticism, our very own share the same perceptions across the ranks.

> *"Camaraderie is a much quoted word in the army, albeit a great strength and part of the professional's pontifical repertoire. But it actually loses steam and virtually stops at the rank of a Brigadier. The General officer becomes more self-centred, remote, callous and indifferent to practicing it. His interest and committal to his colleagues, juniors and the system (to keep it responsive) are markedly on the wane, almost desiccated as he progressively identifies himself with the organisation and status quo which is all so convenient, easy and un-taxing. As he becomes old and tired, and resigns to the environment and the system, he naturally mutates into a piece of furniture losing touch with the ground feeling, but getting into the fashionable "macro-view", "higher perception", etc. Fellow feeling and comradeship take a back seat and even dispensed with, if it conflicts with...."*
>
> *Source: "What ails the Army Officer Class?" By Lt General SC Sardeshpande*, published in www.indiandefence review.com dated 21 Nov 2015.

I commanded a RR sector in the Valley under Gen Bikram Singh as the corps commander. He would blow his top at every untoward incident, despite having experienced an untoward incident of grave nature as the sector commander himself. Conflict, by its very nature, is about chaos and unpleasantness. We do not need commanders in battle who get pissed off at everything and breathe down their subordinates' back. We need commanders who inspire confidence and appreciate ground complexities.

In 2015-16 an incident took place in the Valley, where soldiers on a mobile check post shot at a car which had jumped the check post, despite being signalled to stop. In the incident two civilians died when the car crashed. The Army Commander jumped to blame the troops without even an investigation. Did he realise the implications of his action? What he did in effect, was to disarm the soldiers psychologically and disrobe them of their confidence and initiative and put the troops' lives at risk for the future.

This was followed by a terrorist attack in Uri, where we lost many soldiers including an officer. I see a link between the army commander's action and this consequence, which puts the men in contact in a terrible situation when to press the trigger. In front are the terrorists who are on a suicidal mission, and way behind are our own commanders who inspire no confidence by their ignorance. Kargil was no better and so was Sri Lanka.

The supersession of Lt General Praveen Bakshi in December 2016, surprised the Army no end. Lt General PM Hariz, the Southern Army Commander, another outstanding officer senior to General Bipin Rawat, understood for the first time what collateral damage meant. The government gave all kinds of justification that convinced nobody in the country including the Army.

While, merit should always be the criteria for selection, at this level who decides merit and on what basis? I am a strong believer that all army commanders or all corps commanders are not equally competent. An officer does not become army commander because of comparative competence, experience or intelligence; he becomes one simply because of age factor after becoming a Lt General, and not all Lt Generals get there on merit.

The Army has a system of electing colonels of the regiment from amongst its senior officers. No one has a complaint on this methodology which picks out the best man for the regiment. At one stage, my regiment had two army commanders and none of them was our colonel of the regiment. The regiment, through an electoral system, selected the one best suited for the regiment at that point, disregarding the seniority. Isn't there a lesson here? What about an appropriate electoral college type of system to pick up the best man for the job? It may not be a perfect system; neither is the present one, and surely, who knows their Generals best? The politicians, the bureaucrats or the Army itself?

The senior hierarchy of the army carries immense responsibility on their shoulders. They are at the policymaking levels and what they do makes a lasting impact on the army. The rank and file expects them to be effective. At that level, in my perspective, being effective means ensuring that the troops have the necessary wherewithal for war, ensuring the command is appropriately trained in all ways and being heard and taken note of in the government circles in matters of importance to the Army. The army hierarchy should be able to bring to logical conclusion the pending issues of 6th and 7th Pay Commission, OROP, status and capability development of the Army, etc. The services have seen regular degradation since the 3rd CPC. Isn't it hilarious that the anomalies of the 6th Pay Commission are not addressed and you have many more from the 7th Pay Commission? What does one make out of the effectiveness of the Army hierarchy from this? Any moron will be heard within their command and area of ACR influence.

Lt General is the senior most rank in the army. All PSOs heading the directorates, various branches and institutions are Lt Generals. He is at the level which interacts with the government and civil administration and plays important role in decision making and long-term planning. The rank and file expect these holy men to make a difference to this army. It is sad that I repeatedly hear Lt Generals saying, 'Why can't someone do something about it?' On many occasions, I have questioned these Generals including a vice chief; who is that someone? If an officer of this rank makes this statement, it means two things to me. First, it demonstrates poor self-esteem and lack of confidence in themselves. Likely, they have been like this throughout their service and contributed little of substance. This is very serious. Second, they are happy with their personal achievements and do not care. Either way, it is a shame and indicator of poor self-worth.

I was interacting with a conscientious Lt General who had retired in 2017. When I expressed my views on the senior hierarchy of the Army, he mentioned his interaction with a well-respected army commander in the eighties, who also expressed his helplessness as an army commander.

I heard two army commanders say similar things in 2014 and 2015. Since when have we been so helpless and continue to be so with no efforts to reclaim our legitimacy and voice? Or have we accepted the status quo as a fait accompli? I saw its manifestations during my last tenure in the Army headquarters.

In a democratic country with a disciplined army, resignation is a powerful tool of protest well within the constitutional law, never mind that another dozen are just ready to step into your shoes, even before they are taken off. Haven't we had army commanders who had serious issues with the chief when they were side-stepped from one command to another without a justified reason? Yet, the army commander after some initial representation, chose to serve his full tenure despite being side-stepped. While that is a personal choice, it conveys a wrong message that these fellows will accept anything to serve one more year with all it carries; similar story when General Rawat was elevated over two competent officers to be promoted to the Chief.

Everyone wants someone else to do it.

What ails the senior hierarchy of the Army?

> *During the Second World War, Colonel Darby was offered the command of a regiment with an increase of one grade in rank, but he refused to take it because he wished to stay with the men he had trained. On the same day, General Wedemeyer requested to be reduced to a Colonel so that he could take command of a regiment. I consider these two acts outstanding.*
>
> *—General George Patton, Jr., Patton on Leadership*

Vietnam saw a low point in the US military. In essence, they lay the blame of poor performance in Vietnam on the poor quality of officers, obsessed with career obsession and ticket punching, neglecting the basic tenets of soldiering.

A War College Study on Military Professionalism on Vietnam brought out the following:

> *'Officers of all grades perceive a significant difference between the ideal values and the actual or operative values of the officer Corps. There is also concern that the Army is not taking action to ensure that high ideals are practiced as well as preached... that the Army has generated an environment that rewards relatively insignificant, short term indicators of success, and disregards and discourages the growth of long term qualities of moral and ethical strength on which the future of the Army depends. They are frustrated by the seniors who sacrificed integrity on the altar of personal success. The study found distorted and dishonest reporting of statistics and officer efficiency."*

How does this report apply to the Indian Army? It could easily be a report on the Indian Army.

What are the major shortcomings of the senior officers? Transformation of the officer cadre starts beyond the battalion command. Thankfully, our Army at the cutting-edge level of battalions is sharp and dependable because of its ethos and regimentation, even though a lot needs to be done on the professionalism front.

Careerism. The first folly in our Army lies in the unbridled ambition that goes beyond potential. The single-minded focus of officers in the rank of Brigadier and above is to get two outstanding reports, somehow. This happens at the expense of doing the right things—doing what will be seen and not what is necessary to make your command operationally effective, short term dividends rather than long term perspective, fuelled by short tenures in which to make their mark. Self-interest and personal advancement takes over and professionalism goes out of the window.

Lack of Strategic Perspective and Intellectual Depth: The senior hierarchy in general lacks a strategic perspective and culture and intellectual depth. Majority are shallow in their thought, creativity and application. I have rarely seen fresh thoughts and perspectives in operational planning at the brigade commander level and above. They have the tendency to toe their superiors' line of thought. Check out the operational plans of any formation, you will see the lack of imagination and new thoughts, despite changes in terrain, weapons, technology and enemy capabilities. Masters at putting old wine in new bottle. Majority of the promotions of Maj Generals are based on powerpoint presentations, communication abilities and peacetime event management. Only one in three Major Generals gets exercised and assessed in handling of troops in field for various reasons.

Joint Doctrine of the Indian Armed Forces was released in April 2017. This is an intellectual exercise and a document that would be read internationally. It was publicly ridiculed for its contents and drawbacks. Sadly, reviews by civil and military veteran academia uniformly found it lacking in major ways, including intellectual content and application.

This criticism is not without merit. There is a lesson to learn. I wish to reproduce a para from one review of the doctrine, in the hope that we realise the need to cultivate a culture of intellectualism and dissent and get the right officers up in the ranks, if you want to avoid embarrassments.

'What could possibly explain such a badly written doctrine and does this suggest that the Indian military is incapable of such an intellectual endeavour? Happily, we do have instances of well-written doctrines—the Indian Navy's doctrine titled "Ensuring Secure Seas", released in 2015 represents one such effort. However, that was perhaps because the Navy has invested in its doctrine writing branch and had consulted widely. But the Indian Army and perhaps even the Air Force do not incentivise and develop such expertise. What is required therefore is not minor edits but a complete withdrawal of this document with a revised version to be released later. Not doing so reflects badly on the professionalism and does little justice to the intellectual capability of the Indian military.'

Anit Mukherjee, *Assistant Professor at RSIS, Nanyang Technological University, Singapore, and a non-resident fellow at Brookings India* published in *The Hindu*, 09 May 2017, e-version.

Articles by civilian critics and retired officers give a perspective on our competence, right or wrong. It must be seen as an opportunity for self-correction. It is for our good that we demonstrate the maturity to introspect and appreciate their perspective as a positive feedback and fix the shortcomings where required.

For most senior officers, reading a few pages is a strenuous exercise and even harder to write a few. They have to understand that the higher one goes, the more he is required to know and that will not come without reading.

Lack of Moral Courage. This remains a major weakness. I have cited many examples from my personal experiences where such officers remain mute spectators, or do the bidding of senior commanders knowing fully well that it is not correct. They are too submissive to their superiors and afraid to disagree for fear of annoying them, all with an eye on the report. That is a grave shortcoming as it has serious repercussions. An officer of this rank must have an independent mind and speak his mind.

It is not always that an officer with moral courage gets graded outstanding in this quality, but most others do, for they find superiors of the same feather. Majority lack the moral courage to act against erring or incompetent subordinate commanders for various reasons.

Interestingly, the senior hierarchy is very discerning in picking up the servile and malleable subordinate who will bend backwards to please them at any cost. The ethical ones with moral courage are left alone.

Personal Standards of Perks and Privileges. There is the tendency to set personal standards of probity. Trespasses are justified to oneself under the garb of perks and privileges. Vast majority squeeze the last ounce from the organisation, all in the name of privileges. We only need to take stock of the manpower with retired officers for years and you will know what I mean. Are we a banana army?

Nothing better than the example of US Vice President Joe Biden, who was ready to sell his house for the treatment of his son, only prevented by President Barrack Obama, who came forward to help. His son himself had been the Delaware Attorney General for two years but was unable to afford his treatment.

Further, Joe Biden, when he relinquished office, took a train back home as an ordinary citizen.

President Obama was buying his own grocery a day after he relinquished the office of the president.

This is integrity.

In India, our very own president A.P.J. Abdul Kalam, the most respected President of the country, set a glowing example. His assets at the time of death were some clothes and lots of books. As the president, he paid for everything he used in personal capacity and for his village folks whom he had invited to the Rashtrapati Bhawan. He cancelled the ritual Iftar party for the rich and famous at the Rashtrapati Bhawan, budgeted at about twenty-two lakhs, added one lakh from his personal money and had food and necessities distributed to orphanages. He wanted to pay for the stay of his brother for a day who shared his room, which his staff prevented him from doing.

Compare this with what we do and it is all covered under the fig leaf of perks and privileges.

Moral lecture? What did I do in service. I made a very serious effort to be correct.

Gen Bikram Singh, then a corps commander at Srinagar, visited my sector in Lolab with General Thamburaj, the then Vice-Chief, who had raised this sector a long time ago. The Vice-Chief recounted his days of lack of resources and hardships and how his stylish successor had then done up things later.

General Bikram said looking at me, "He is also from a stylish regiment and he must have also done up his room well."

I replied, looking straight in his eyes, "Sir, pardon me. We have done a lot here in the sector as you can see. But there are three places where I have not spent one penny-my accommodation, my office and my vehicle." The same was the story as a commanding officer and a divisional commander. As a commanding officer and sector commander, there was no AC or heater in my vehicle or office, simply because my men used to be out in the freezing cold everyday. As a sector commander in the Valley, I chose not to claim transport allowance, though legal as per rules, but unethical in my perspective.

As a battalion commander, working came to my house only twice. Once when I unloaded my truck on arrival, and then when I was posted out two and a half years later. My wife drove in our car for her classes.

The commanding officer of 9 Infantry Division, Engineer Regiment, Col Bakshi, was the Golf Captain of Meerut Golf Course, the station where I commanded my division too. Later, when we both were posted in New Delhi, he said in a gathering, "General Sengar was the only General in Meerut (there were five), who paid his caddie money on the golf course." That indeed came as a surprise to me. Is there any other way?

If this is not enough, no funds whatsoever, either the MES funds or regimental funds were spent on the Flag Staff House as the divisional commander or when I was the Deputy GOC in Nagaland. As the Deputy GOC of 56 Infantry Division under raising, my entire staff comprised one sahayak, one office runner and one driver.

While my division was partly located in Meerut and partly in New Delhi, my children never travelled by army vehicles or CHTs. They either drove in my private car or travelled by bus, to the surprise of the commanders and the staff. The same for my wife, except when travelling with me.

Professional Mediocrity. The present system allows for average officers to scale the highest ranks through being 'system-fit', meaning managing the environment.

Regimental Parochialism. Promoting their cronies based on regimental, caste and other affiliations at the cost of more deserving officers.

Absence of Role Models. Very few officers earn the genuine trust and respect of their command.

Lt Gen A.K. Singh, former Army Commander Southern Command and the Governor of Andaman and Nicobar, categorised the senior officers as follows:

- A small percentage of outstanding military leaders, with clarity and conviction, well-versed in the higher direction of war.
- Fewer still qualify as role models, based on integrity, competence and commitment.

- A percentage of mediocrity manages to get in—not being able to rise above the tactical level.
- Status quo approach and turf considerations manage to retain hold of many at the cost of the organisation.
- Shortfall in visionary leadership to measure up to the 21st century challenges that becomes apparent.

(Talk at Lt General Hanut Singh Memorial Lecture, 9 May 2016.)

He makes a good point that good leaders exist in our midst, but they need to be recognised, nurtured and supported. The present system seems to beat the man with these qualities that will win battles.

I was shown an informal survey by a very responsible Maj General, conducted on the senior army leadership in the Higher Command Course and the College of Defence Management. While sharing the statistics in this book will be embarrassing, it is revealing what the middle rung officers think of the senior hierarchy as far as integrity and leadership is concerned. These studies, separated over many years, revealed the same impression. The Commandant Infantry School shared the findings of his informal survey conducted amongst the junior officers in the rank of Lieutenants and Captains. The faith expressed by them in the senior officers of the Army was very, very discouraging, to say the least.

Where does the fault lie? It lies in the promotion policy and the weak fabric of senior leadership that promotes officers of doubtful integrity. It lies in the defunct value of confidential report, on which all promotions are based, as a document to assess the character traits of star rank officers. Infantry, holding the most important policymaking positions, is paranoid about losing vacancies to other combat arms, and finds ways to scuttle scores of representations made by other arms.

I have dealt with these aspects in the chapter on Promotion Policy. Just to repeat, why should an officer graded outstanding in integrity by over forty officers in the last two dozen confidential reports, decorated twice for distinguished service, promoted to the last pedestal of Lt General rank, have to be court martialled for corruption? Why should the service chiefs be in the dock for impropriety, if your character assessment in the confidential reports was accurate and true?

While there are good officers in plenty, they are still not the majority, as they should be in the miniscule percentage in these ranks. As late as April 2017, the general impression by the middle rung officers, being aired on the social media, was less than flattering. What does that say?

- Disconnect with middle rung officers, socially and professionally.
- Shades of grey in rights and privileges, leading to rampant corruption.
- Lack vision/insight and are too self-centred.
- Professionally dishonest and promoting mediocrity and a culture of sycophancy.
- No transparency, dynamism or boldness in decision-making.
- Fond of flaunting badges of rank and privileges.
- Inconsistent policies, so the system continues to be personality oriented.

2017 saw multiple terrorist actions including the beheading of our soldiers. During the elections in the Valley in Apr/May 2017, there was also an incident of a civilian tied up in front of an army vehicle by an Army rescue team to rescue police and polling staff from almost 900 stone pelting marauders vying for their blood. It was a brilliant out of the box innovation that saved many lives. A number of serving and retired Lt Generals appeared on the TV to condemn this officer, giving lecture on morality and ethics, without recommending what he should have done.

This is what Col Rajender (Retd) wrote. Excerpts reproduced from the social media:

> *"India: A Nation of Cry-Babies*
>
> *India is led by cry-baby politicians and her army is led by cry-baby Generals—both just talk.*
>
> *Stone pelting has made our soldiers nincompoop and our Generals harangue them on the narratives of ethics, if they react with innovation.*
>
> *Do our Generals have it in them? I am reminded of an old saying, 'I am not afraid of an army of hundred tigers led by a sheep. I am afraid of an army of hundred sheep led by a tiger.'*
>
> *This is the crux of the problem.*
>
> *We have all sheep at the top of the ladder, who keep issuing statements to hide their incompetence. We cannot do anything because the weak and the inefficient have risen to be Generals.*
>
> *Remember Gresham's law of economics—'Bad money drives good money out of circulation.' Similarly, in army, careerism has replaced good officers of the army with bad officers. This is the harsh reality."*

General Bipin Rawat restored confidence by awarding Maj Gogoi the Army Chief's Commendation. That is good Generalship.

Why does this impression prevail in the Army? Decades apart, nothing has changed in the perception of the middle rung officers. Are we so blind that we cannot see the writing on the wall? We have the formation commanders in direct contact with officers and men. We have the Public Information Directorate, whose business it is to monitor the pulse through keeping a tag on social media and we have the military intelligence. Or have we crossed the point of no return? As Mohan Guruswamy says—Flawed Generals, who will bell the cat?

How different is this perception, in its sense, from what Lt General A.K. Singh has said in his categorisation of senior officers? Not very much.

Well, one can find many faults with this exercise, or blame both the middle rung and the junior rung for the shortcomings that they may have, but you cannot wish away the fact that it reveals a lack of confidence amongst the vast majority, in the vast majority of the senior leadership.

You can condemn these the way you wish. It is the perception of the majority officers and demonstrates lack of 'trust and confidence' in senior Army leadership, and it matters. It is time to sit up and take note.

The net assessment of General A.K. Singh's views is that we are far short of leaders who have what it takes to take the Army forward in this century.

During my last two years' service, I served in the Army headquarters and watched the functioning closely. The general impression that I have developed is that officers in the rank of Lt Generals and other senior ranks, while doing their narrow domain tasks, are wary of taking decisions, shy of raising difficult questions and tough feedback and make no efforts to make a difference in issues of larger interest. They are happy to coast along, contented in their own achievements, with an eye on a PVSM. The military bureaucracy is competing with civilian bureaucracy in avoiding making decisions of substance. Playing safe suits all.

I attended two army commanders' conferences. Sadly, I could perceive a sense of helplessness in basic issues such as rations, infrastructure development, financial powers, development on pay commissions, OROP, etc., and on decisions taken by the MOD with far-reaching effects on the functioning of the Army without consulting them. My impression is that the army commanders have the powers to redress most situations, but many lack the courage to exercise their powers out of some fear.

Going by the media reports, General Bikram Singh and General Dalbir Singh, who were Eastern Army Commanders, carried a reputation of signing nothing of significance for fear of putting their appointment to chief in jeopardy. According to the *Times of India* dated 14 Jan 2017, they averaged an expenditure of 3 to 4 crores in their tenure against the allocation of approximately 100 crores, as against 100 crores by Lt General Praveen Bakshi, which needs to be lauded rather than be investigated, unless there is corruption involved. The net sufferer is the nation and the poor state of security infrastructure in the North-East is not lost on anyone.

Lt Gen H.S. Panag has suggested the following competency requirements and qualification for the post of the Chief:

- Capacity to provide best advice to the RM, PM and the Cabinet Committee for Security with candour and forthrightness and more importantly, the moral courage to disagree when required, before a final decision is taken.
- A wide-ranging intellect with broad knowledge of strategic and military affairs, understanding of the constitution and government functioning beyond military matters, and backed by a broad based varied command, staff and operational experience as a higher formation commander (division and above).
- Ability and will to execute the National Security strategy and Transformation Strategy; create and sustain the deterrent decided by the government; and to lead intellectually and managerially his subordinate commanders for the same. Tolerance for complexity and ambiguity.
- A man of character with impeccable integrity, moral courage, communication skills and bearing with a strong sense of ethics and trustworthiness—a role model for his service and the country.
- At international fora and in military diplomacy, his personality must be worthy of admiration. The adversaries must hold him in awe.
- Believe in civil government supremacy, have credibility with the public, media and parliament, and compatibility with the RM and the PM.

Quoted by Maj Gen PK Mullick, VSM in his article, "Selection of the Chief".

I would agree with that entirely. I would like to add that the sense of these qualities and skills are as much applicable to all Generals and the army commanders. The question is how do you ensure that the senior hierarchy, from which the chief emerges, acquires these traits, knowledge, wisdom and skills. These would not appear suddenly from somewhere at that rank. There has to be a culture of scholar warriors-intellectual development, creative thinking, development of operational art and operational logic, development of risk-taking abilities, tolerance for ambiguity, the vision to see beyond the fog and uncertainties of war, etc., which is a weakness in senior Army officers, as a majority are comfortable at the tactical level with few fresh ideas. Higher direction of war and operational art are all about defeating the adversary in his mind before the first shot is fired. There is no substitute to knowledge and application and that will not come without protracted intellectual growth. As long as it is not a necessity for promotion, very few will do it of their own will.

In his lecture, Gen A.K. Singh summarises the expectations from the senior hierarchy as follows:

- Transformational leaders with a high tolerance for ambiguity and professional dissent.
- Have a leadership vision with a developed sense of battlefield intuition.
- Victorious will with calmness in crisis.
- Be able to co-relate the external influences with internal environment and without compromising the core values; shape the environment to optimise the end state.
- Army's General will be expected to expand, not constrain, the nation's range of strategic options.

The views of Gen Panag and Gen A.K. Singh, clearly bring out the expectations and capabilities. An honest introspection leaves no doubt of the voids and shortcomings, which are blindingly glaring. The issue now is what needs to be done to face the challenges of the future.

National security is not the Army's baby alone. The CCS, the PMO, the Defence Minister and at the functional level, the MOD bureaucracy has to be held accountable. The Secretary of defence is responsible for the security of the country and not the service chiefs. The Army must ask him and the Defence Minister questions when serious issues are not addressed rather than plead.

Sadly, one army commander once mentioned, "You guys think that the army commander is a big man. No, he is just another man."

What do I infer from this? I infer that the army commanders do not see themselves as people who are taken seriously or heard and can make a difference.

That is a serious situation because that is the highest rung of leadership. They are expected to carry enough clout to be heard in the corridors of power of the Ministry of Defence and the government where it affects national security and the Army's interest. If their area of influence only coincides with their ACR domain, then it is a shame, which seems to be the case.

Have the army commanders become redundant in this context?

During the Army Commanders' Conference, there is a session called MOD interaction. All the joint secretaries headed by the Defence Secretary attend this session. It is meant to focus on long pending cases that have not seen the light of the day due to MOD bureaucratic lethargy. Most of the cases that are brought there have been languishing in the MOD for over a decade. I am given to understand that the interaction agenda is screened by the MOD. What is embarrassing due to inaction at the MOD is removed from the list. If the Army has included an agenda in the list for the MOD interaction, it is obviously critical. And the Army must insist that the item cannot be removed from the list unless it is brought to satisfactory conclusion before the event. In my view, from what I witnessed, the current methodology of MOD interaction is a complete waste. What is being discussed today are individual issues, whereas, what must be streamlined are the policy and procedural issues that makes functioning effective, expeditious and accountable.

Where the Army must take the blame is that it allows this rare MOD interaction to become just another event, rather than seeking accountability for inaction. I got a clear impression that the Army is almost pleading for its long pending cases, in some cases over twelve years, to be expedited, as if it was asking for a favour. I was aghast at the submissive tone of officers, Brigadiers and Lt Generals who spoke before me. I had to present one case relating to a simple case of authorisation of tank T 90 ammunition, languishing for twelve long years. Precedence of authorisation of tank T 72 ammunition existed, duly approved by the government, as such it should have been resolved without too much fuss. I made it clear addressing the Defence Secretary, that the issues that we discuss here are neither an individual directorate issues, nor army issues. They all affect the country's security and therefore, are the country's issues. I said, in my case, the file was with the JS(ON) for over twelve months on two occasions with no action and no accountability. I stated, I am not worried about my case. What I want to question is the systemic problem of delays, lack of urgency and total lack of accountability of officers in the MOD for inaction. How has that joint secretary been held accountable for sitting on the file for the whole year, without justification? I mentioned a few other examples. After the session was over, a Lt General, who also presented a case that day walked up to me and congratulated me, saying, you sorted them out today. I told him, don't congratulate me, you also spoke, why didn't you say so? These holy men in the higher ranks enjoy immense perks and resources of the Army and seem too timid to speak out and wait for someone else to bell the cat.

In the same conference, I was surprised and disappointed by the submissive manner in which the army commanders were talking to the Defence Secretary, one of them addressing him as sir, six times in a minute. The manner of addressing these civilians is clearly laid down which is Mr X or by his appointment. I have seen Lt Generals salute the Defence Secretary and secretaries. This is how we have eroded our credibility where it matters. When we do this, we set a very wrong precedence, which gives these bureaucrats an inflated sense of their position that does grave damage to the Army. I remember reading a letter by our first commander in chief, Field Marshal Cariappa, who said that when we address the civilians as sir, for personal gains, small or big, we do grave damage to the Army. The government has smartly placed a carrot in front of every rank. The chief is looking to be a governor or an ambassador, army commanders are looking to be appointed in the Armed Forces Tribunal or government assignments, and Lt Generals are looking for a PVSM, for its monetary benefits or some placement. They need to remember, if they have reached that rank even by default, they have an obligation to the Army. I have found it most silly when a PSO or a Lt General says, why cannot someone do something about it, or the system is like that. For God's sake, who is the system, if the Lt Generals are not?

To the chiefs looking to be ambassadors or governors, I would make a plea. Most of these ambassadors are of the joint secretary level. Please do not lower the high esteem of the office of the Chief. The whole Army looks up to that office. For those Lt Generals eyeing the PVSM, please realise: one-It does not make you any taller. Most see it as a vulgar dilution of the award given to the majority in that rank for doing the routine, and not for any achievement. Two-the financial benefits of the award makes no difference to one's life. How often have I heard senior officers referring to the PVSM as a parting gift and that says it all.

So often I have heard the senior Army hierarchy being described as 'spineless' by the middle rung officers, which will take no decisions, that I can't help believing that this is the case. Everyone plays safe, happy in his own achievement and personal growth. My experience during the service and as the divisional commander, which I have elucidated in the chapter on command assignments, very strongly makes me believe that is the case with the majority.

Lt Generals are expected to put high value on themselves. If your area of influence is restricted to your ACR domain, you are redundant. You are expected to make a difference to this Army with your stature and credibility to be heard beyond the Army.

General Handoo was the Army Commander in 1983–84. He was one of the most charismatic Indian Generals who is said to have left the Prime Minister's Office because he was kept waiting after having been given an appointment. He told the PS to the PM, there are far too important things happening in the Northern Command for me to be wasting my time waiting here.

They placed a value on their rank and position and when they talked, the people in the government listened. They were men of great stature, character and self-belief. We want our senior hierarchy to be able to alter the status quo, force the pace at which important things are happening and re-write the narrative where required in the best interest of the organisation and be heard in the corridors of the MOD and PMO.

Interaction with the Minister of State (MOS) for Defence

I was officiating as the DG of Mechanised Forces Directorate for almost one year in 2015. One day I was called by the MOS for Defence. The personal secretary said there is no agenda; he just wants to discuss a few things. So, I landed without a brief.

Amongst other things related to equipment of my directorate, he asked for my suggestions to improve the functioning in the MOD. I said, apart from known problems, the biggest problem in the MOD is the lack of urgency and accountability amongst the bureaucrats for the time taken in processing files. I quoted two incidents in which a file of my case was kept with a joint secretary for over a year without acting on it.

So, I suggested some measures to ensure that the MOD officers are time bound to act on a file, failing which a delay report must accompany the file. List of all delay report cases with the details of the officer responsible must be put up to the MOS for Defence/Defence Minister through the Defence Secretary. They should at random inquire into these cases.

He agreed with me entirely, but obviously did nothing in the end. He narrated a case where PM Dr Manmohan Singh, initiated an effort to make the bureaucracy accountable through periodic performance reviews. He said, these bureaucrats kept the file in a spin for five years, ensuring nothing came out of it and the government changed. Is everyone helpless against the bureaucracy in this country?

The story is no different in equipment procurement. The first hurdle is the military bureaucracy, where new incumbents feel that their predecessors were useless and start reinventing the wheel, only to realise what was done was a well deliberated institutional decision, putting the projects back by years. I dealt with the Future Infantry Combat Vehicle (FICV) and the Future Ready Combat Vehicle (FRCV—next generation tank). All our efforts and approvals of the Defence Secretary have been thrown into the dustbin by simple, unquestionable CYA (Cover Your Ass, in the military parlance) comment—follow the Defence Procurement Policy (DPP) 2016 by the deputy chief of that time. We have seen, how much has happened through the DPP.

In the MOD, there is no single point mechanism to coordinate, expedite and push the three pillars of a project in the MOD—the Secretary Defence Production, DG Acquisition and Secretary Finance—towards the success of a project. Each is concerned by only covering his tracks, becoming slave to the procedure and rules, doing the CYA, with no concern to what happens to the project. As a result, conscientious mid-rung army officers are running from pillar to post. Nothing concrete has happened in this regard and despite the fanfare with which the DPP 2016 was launched, little will happen unless there is paradigm shift in the MOD basics. People have to be held accountable for why things are not happening, rather than hiding behind procedures. What has happened to the FICV project, when the evaluation report was handed over to the MOD around November 2016 after a magnificent effort by the Mechanised Forces Directorate?

The senior Army hierarchy is content to let things happen at their own pace, making little efforts to expedite things in the MOD even for these critical projects.

I have narrated some experiences right through my career where I was personally involved, or was close to the events to certify their authenticity. The number of incidents that I have personally experienced leaves me in little doubt that it is the norm with the majority. At the end of this, I would like to ask this question,

'How professional is our hierarchy?'

I have only narrated some of my experiences at different stages in the book. Sri Lanka and Kargil only reinforce my impressions. Simple extrapolation means there must be numerous instances of such display of professional and moral bankruptcy happening elsewhere. In all the incidents that I have narrated, a total lack of character is on display. Commanders wanting to give a birthday present to a senior commander by capturing a most useless piece of ground, which would only be a perpetual headache without dividends. Yet, such incompetent people rise to higher ranks in the Army.

All these only support the views of Mohan Guruswamy; hierarchy which is 'Visibly dense and obviously incompetent.'

I had expressed my fears a few years ago, that given the way the Army is handling its promotions, soon the government will step in and that seems to be happening. Looks like we may be arriving there. The biggest damage of this exercise is the loss of credibility of the Army as an institution.

Our most precious commodity as a soldiering community is our pride, honour and trust. Let us not lose them by the smallness of our minds.

Decorations and Gallantry Awards

Abbreviations—Sena Medal (SM), Vishisht Seva Medal (VSM), Ati-Vishist Seva Medal (AVSM), Param Vishisht Seva Medal (PVSM), Yudh Seva Medal (YSM), Uttam Yudh Seva Medal (YUSM)

Awards are meant for exceptional contribution beyond the call of duty and not for doing what you are meant to do. Our Army gives lots of awards for doing the routine. If one sees the never-ending list of awards given by the Army every year, the Indian Army should be full of role models. While the gallantry awards earned in operations, not always commensurate to the action, are acceptable as a recognition and motivational tool, the trouble is with the so-called awards for distinguished service of Vishisht Seva Medal (VSM) series and the Sena Medal (SM) (Distinguished) given in the Army headquarters to officers of the rank of brigadiers and above and personal staff officers of senior officers. The long list of decorations that so many officers carry, without having been in the line of fire or even being in the field or having achieved anything of substance, to me, sounds painfully vulgar.

Nearly every officer commanding a division in field is awarded an AVSM or YSM, and they are all, barring exceptions, infantry officers. Every corps commander in field is awarded a UYSM or a PVSM for doing nothing of substance. Those who command in peace, known to the chief, may get one too.

I was told by a General officer who was deputy military secretary to one of the army chiefs, that AVSMs are awarded even without citations on the whims and fancies of not only the chief, but at the whims and fancies of the deputy military secretary himself. Sometimes, it is so vulgar that one wonders what these senior officers are made of. A few years ago, the staff officer of the Military Secretary, in the Army headquarters, was awarded SM. The General himself was court martialled later in a scam. Colonels galore in the MS Branch were awarded VSM when decorations carried weightage for promotion in the quantified system. All Brigadiers in the WE directorate were awarded Sena Medals in 2015 and we know where our equipment procurement is. In 2015–16, the brother of the army chief was awarded VSM one year and SM the second year as the commandant of the army camp in Delhi. His unit itself was awarded the COAS unit citation. What about the rest of the hardworking officers?

No one expects the chiefs to be Harish Chandra, but we do expect them to act with minimum discretion. The other people, equally to blame are those in the chain of award, from his superior officer recommending, to the awards committee.

Today, there is no dearth of Generals who have at least four decorations, PVSM/UYSM, AVSM, SM, VSM, etc. Now compare this with Field Marshal Cariappa's calibre, character and contribution, who shouldered great responsibilities as the first commander-in-chief, was only OBE. General K.V. Krishna Rao, the former chief, was also the governor of J&K in very turbulent times. His decoration when he died was the solitary PVSM, and this was awarded for the command of a division during the Indo-Pak war 1971. There was insurgency in the North-East and the challenges on the Line of Control that time too. Are the contributions of the present multi-decorated Generals many times over those of the Chiefs? It impresses no one and sounds so hollow.

Many of these highly decorated officers were court martialled for lack of integrity. Does it not ring a bell that the whole system needs a review?

The second malaise is the chief's and the army commander's commendation cards. The largest beneficiaries of these awards are the personal staff officers, regimental officers and their personal staff-drivers, sahayaks, waiters, runners, etc.

I was narrated this revealing story by one Deputy MS to a former Chief.

A Major General, waiting in his office to meet the Chief said to him:

"I admire this Chief. It would be an honour to have a COAS commendation card from him." Since he was in an appointment which enabled him regular interaction with the Chief, he said that a couple of times over a period of time. The Deputy MS took out a Chief's Commendation badge from his drawer and gave it to him. And he was wearing it the next time when he came to his office! That man went on to become the Chief. Sad, such people find their way to the highest office. This officer also told me that he gave three COAS commendation cards to a JCO on his request in six months. He was PA to this chief over many tenures and these would help him to get honorary commission. And all these without a citation. Isn't that a cruel joke?

I was the Deputy GOC of 56 Infantry Division on raising. Two brigade commanders and three divisional unit commanding officers raised their formations and units in Nagaland/Arunachal from a scratch in the most challenging circumstances and took over operational responsibility in record nine months. We recommended them for the award of VSM, which I was convinced was well deserved by any standards. Net result, not one was awarded even a COAS commendation card, whereas people sitting in the Army headquarters in Delhi and personal staff officers continue to be awarded VSMs and SMs. A very senior officer serving today was awarded an SM and a VSM while posted to the MS Branch. Look at the mess we are in, in MS matters, where every odd officer feels he can get justice only from the court.

Now consider, how do you compare a COAS commendation awarded to an individual, against a citation for a Sena Medal for gallantry, with one given to personal staff officers even without a citation? To cut the story short, the senior officers have permitted the decorations to be diluted to a joke. Neither is the recipient proud of it, nor anyone else. I have seen officers awarded such decorations feeling too sheepish to wear them till they get posted out.

In 2013–14, the divisional commanders in Hissar and Dehradun were the only formation commanders in peace to be awarded AVSM. It may just be a coincidence that they were both infantry officers and belonged to the same community as the Chief. What about the balance three dozen divisions in peace? Did those divisional commanders have a holiday? When I was a Colonel posted in the headquarter at Ambala in 2006–2007, a divisional commander in Patiala, of the same community as the then Chief, was awarded VSM and AVSM in two successive years. In 2015–2016, only two Major Generals in Western Command were awarded VSM; they both belonged to the same arm as the army commander. Such incidents do not go unnoticed and chip away at the credibility of the senior officers. Our senior officers seem not to grow beyond narrow regimental parochialism. The social media is a perfect tool to get a feedback on where we stand, but we see no evil and hear no evil.

I remember the statistics of two years, where more than 90 General officers were awarded these decorations, that is, more than thirty percent of the total strength of the Generals in the Army. Some of them later were in sharp focus for corruption.

What needs to be done?

The Army should review its policy on awards. They have been diluted to a state of vulgarity. For genuinely outstanding contributions on staff, a letter of appreciation may be considered by appropriate authority to be published in suitable gazette. There should be no medal decorations for personal staff officers. It will avoid embarrassment to the Army and restore the value of awards won in operations. SM (Distinguished) should be discontinued in non-operational areas and in staff. Clear parameters of achievements should be laid down for awards in command to remove discretion. Length of service and routine work is no criteria for an award. One wonders, how is the awards committee sitting in New Delhi qualified to make the final judgement, even though there is a need to screen the inputs from the initiating agencies. The responsibility lies with the senior commanders at the initiating point to ensure that there is no dilution. Sadly, that is the sense which is missing in most.

In 2016, two Major Generals, one from the ASC and one from the Ordnance Corps, were approved for promotion to Lt General out of over thirty Major Generals. Clearly, they must be having a near-perfect record of service to be three out of thirty in a group where everyone has an outstanding record. They were both well decorated. I am given to understand that they were also in line for the award of PVSM. The Army was saved some blushes when they were placed under CBI enquiry for corruption at the right time. These are not isolated cases; these are the tip of the iceberg that have been exposed. Ask their colleagues and their spoken reputation is laid bare which is farthest from flattering. One wonders how the wise men of the Indian Army can be so ignorant? It is clear that their superiors were in the same mould and were incapable of taking these officers to task, either by choice or fear of own vulnerability.

This brings me to the important issue of a 360-degree input on the senior officers which I will deliberate on later.

Why not consider discontinuing distinguished award for star-rank officers? Look at FM Cariappa and Gen Krishna Rao. Do these officers, supposed to be the cream and highly motivated, need incentives? At least you will eliminate acrimony and perceptions on the fairness of awards.

The other area is the award of Unit Citations. An analysis of the awards and the regimental associations of senior commanders in the command chain will reveal the malaise. Can there be pro-rata in the award of citations based on arms? Surely NO. Unit citation in peace is akin to battle honour in war.

It must be earned by attaining the highest standards on all counts, especially operational readiness demonstrated consistently over a period. We need to look dispassionately at this aspect as well.

The bottom line is, in everything in the Army, merit must be the only criteria.

> *"Once we start honouring people for doing what they are supposed to do, the organisation will certainly start downward slide. We must reward for the extraordinary, beyond the call of duty... In 1980, I was one of the 87 COAS commendation winners. Today up to 1000 are awarded in a year!! Times have changed but none should be made to feel that one can crook his way to a star rank. Solution again lies in ethical leadership. The way to begin is we make a change wherever we can and initiate changes in the minds of those who can change things."*
>
> *—Lt Gen KH Singh (Retd)*

* * * * *

WOMEN AND THE ARMY

Women in the army has been a controversial issue. The Chief, General Bipin Rawat spoke of inducting women in the Corps of Military Police. In my opinion it is not a good idea and the serving officers will not tell him. It will set a precedence that I believe is not in the operational interest of the army. The politicians, the women rights activists and the courts have all been advocating women's induction in all branches of the army. Those who have not worn the uniform will never understand the nuances. Eventually, the organisation will learn to live with whatever is imposed on it, but there will be a price, and the price will be paid by their male counterparts.

Quoting European and US examples is comparing apples and oranges. Their reasons for having women in their military are different from ours as are their culture and operating environment.

While no one questions their ability to handle most of the peacetime routine responsibilities, the army is not about peace. It is about the blood, shit and grime of combat. Combat can be demanding and brutal. For those who fall into the enemy hands, Pakis or the terrorists, it can be terribly, terribly inhuman. We have seen that in the Kargil War and in the Valley. God forbid if a woman was to fall into their hands.

Even in peace time, it involves full time, round-the-clock commitment where your personal affairs always take a back seat. In the operational environment in the Valley and the North-East, lady doctors, who are a necessity, not for the soldiers but for their families, do not shoulder the risky and tough assignments as RMOs with the units because of the vagaries of the operational environment and their vulnerabilities.

Commanding 8 Sector RR, I met an army doctor on attachment with one of my units. He told me, out of his two years in the Valley, he was out of his unit on attachment for almost twenty months. The reason the lady medical officers were not sent on risky duties from the comforts of field hospitals was because of the nature of the duty and safety concerns.

So, while they may be good doctors, who will do the dirty work? Who will do the long-range patrolling with twenty kilos on your back? Who will go on three days operations up in the jungles and the mountains with the battalions? In the early part of their service, we need doctors who are also soldiers. Later, they can be doctors alone. The lady army doctors fall short in operational environment that the army operates in, so do the majority lady officers of other services once the family responsibilities come in.

On two occasions, I had lady officers under me. In both cases, their responsibilities as mothers came in the way of discharging their duties. One was seeking a cancellation of posting to a field area after four Class A city postings. The other was seeking favours from everywhere to avoid going for the six-week exercise because she had a small child and the husband was in field. While one might say that one needs to be compassionate, that is exactly the point. They have very demanding domestic responsibilities to be able to meet the demands of an operational military. A couple where both serve in the army, means difficulties for the couple itself and compromise for the army. There may be women CEOs, but corporates are not armies. Exercise in peace is equivalent to war in war. Who will do their job?

The Army cannot afford to have a class of officers and soldiers who become non-operational for long because of personal obligations and gender compulsions. The operations and the shortage of officers do not provide you that luxury.

The problem with ladies as soldiers in the army lies on many fronts. First and foremost is their domestic responsibilities as mothers and housemakers. A couple where both are in professions where one does not have the freedom of absence from office at critical times means trouble. And a majority get married to service officers.

The second is the physical disadvantage which is biological. It is not about running and pull-ups. Surely there are women climbing Mount Everest. But that is not soldiering. There is so much more.

A young officer in the Engineer Regiment is the hardest working recruit, jumping into canals and rivers, rubbing shoulder to shoulder with men lifting heavy bridging equipment, loading vehicles, laying mines etc. I have not seen lady officers do that. That is one pair of hands less on deck that could be critical sometimes. Same is the case with Corps of Signals, Air Defence and

Artillery. It is doing all this, that young officers build that camaraderie and earn the men's trust. You do not earn their confidence standing on the side when they are sweating and slogging.

To accommodate them in office-bound jobs, as often happens today, is to deny the male officers the much deserved break and that means compromise. Partly, we are to blame for not demanding what they have signed up for.

The third aspect is cultural. We are still some distance in this regard where women will be comfortable with the men the way we live and serve as brothers-in-arms, particularly in an operational environment. Women's presence increases security and logistics burden on the units. It is not about male chauvinism, it is about what makes the Army more effective. Anything that weakens the Army in operations is most undesirable. I do hope that a thorough examination of the value addition that they bring to the CMP in peace and war, their employment across the vast range of deployment and the impact on additional load on the men soldiers has been carried out.

Posted as Colonel GS in a corps headquarters, the staff officer to the corps commander told me that the lady officer under me has sought the corps commander's interview against me, stating that I am not allowing her the freedom to work freely. He asked whether I would like to resolve the issue at my level or he should take the matter to the corps commander.

I said, "I have never in my career done anything that I am afraid of or have to hide. Give her the corps commander's interview." The corps commander felt it was not an issue to be brought to him and delegated it to the chief of staff, who in turn delegated it to the BGS and he to me.

When I told this lady officer that she had taken the corps commander's interview against me, she shot back, "Yes, he is my commanding officer. So what is wrong?"

I said, "You can take the interview of the Army Chief, but the request had to be processed through me, as I am your commanding officer and not the corps commander."

After a few days, I called her on phone at about 8.30 AM. She retorted, "Are you calling to check if I have come to the office on time?"

Similar thing happened again. When I called her up at about 1.30 pm, she retorted that was I trying to check if she is still in the office? One day, when I rang her up for some work, I was shocked to learn that she had got herself posted out to the Engineers Branch in the same headquarters, and I had no information. So much for the MS Branch in the army headquarters.

In her report I reflected her shortcomings and recommended that she not be given extension. That shook her up. She had been pampered for too long. Soon, she chose to put up her papers.

I had an interesting experience as a commanding officer deployed and ready to go to war during Op Parakram. I noticed a largish tented facility come up suddenly, much bigger than mine with a screen covering it from all four sides. I assumed it must be some Brigadier who has moved in there. I was told it was the accommodation of a young lady officer posted to the unit workshop. Much against my representation, they still posted her to me.

The next day interview was interesting. I welcomed her to my unit and asked her if there was something that I could do to help her settle down quickly. She told me her sob story of how her parents were not agreeing to her getting married to one officer. And her strange request! Could I perform her father's role and oversee her marriage in a temple at the earliest? Well, I said we will sort out Pakistan first and her marriage will have to wait till the war is won.

While they may sign any contract, when it comes to the crunch, their domestic responsibilities and challenges as women make it extremely difficult for most to measure up to the challenges. These and all kinds of leave of absence authorised to women, makes it too disruptive at the officer level that the army can ill-afford. The Army is not the place for gender equality. Let the men do the fighting. The others will not understand this. The army senior hierarchy must put this across convincingly and emphatically.

* * * * *

The Parting Shot from a Soldier's Pen

Dear Colonel,

I superannuate on the 31st Jan this year. I have loved being a Guardsman every inch of the way and enjoyed the high expectations that the environment has from a Guardsman (officers and men) and Guards Battalions. I write with the hope, that as the Commanding Officer of a Guards battalion and beyond, you will nurture this spirit in all your men and leave this letter to be handed over on relief, if you find it appropriate.

This brings to an end an extremely satisfying and purposeful journey as a Guardsman. Sadly and inevitably, it ends my active association with the profession I loved and gave it all. I shall miss the Garuda lanyard and the brass buttons. I shall miss the spirit of the 'Red and Gold'. I shall miss the splendid 'Non-Transactional Company' based on mutual trust, this profession provides.

I have always carried an image of a Guardsman in my mind. I would like to see them to be proud officers, have an immaculate bearing and be well turned out at all times, be professionally outstanding and troop-oriented and live by military values. I would like them to be thorough professionals who live an honest Spartan life and sleep and live on air and water. They must create ripples by their presence through their professionalism, moral and physical courage, simple living and high thinking.

I like to see them as proud ambassadors of this wonderful army.

I like to see our men develop in the same mould. They must make the heads turn with their pride and their gait as Guardsmen.

The ladies must carry the same ownership and spirit of the Regiment and play their part in bringing up the regimental team spirit.

For a unit, there is no Fixed Deposit of good reputation on which you can dig in crisis. You will need to add to your 'Reputation Account' every single day. All this would not happen, unless you as the Commanding Officer, walk the talk and set the example. There are no different standards for the Commanding Officer. In all cases the Commanding Officer eats last.

Remember "Command is a payback time—an opportunity to give and not to take." Leave your legacy and that will happen only when you live selflessly—place the interest of your battalion and your men before your own.

It will always be great delight to see the Guardsmen and units scale the heights of professional excellence through hard work and fair play.

I wish the battalion good health, happiness and all success.

Jai Garud,

Maj General Anil Sengar

Jan 2016.

CONCLUSION

Life in the army has been a glorious journey. There were challenging times as well-glories and disappointments, but my enthusiasm never waned, because it was based on foundation of a principled living and military values. I cultivated a great attitude; always saw my blessings and appreciated what I had, rather than brood over what I did not. I was never unduly elated at my accomplishments, neither too disappointed at my failures.

I never grudged anyone success and happiness, even though I may have had opinions. If we see the balance sheet of our lives, the credit side far outweighs the debit side. Most of us lack the 'attitude of gratitude' and choose to see what we do not have, rather than what we have—a recipe for unhappiness.

I believed that I did not have to be serious to do my work seriously and developed the art of loving the work that fell in my lap. That was the secret of why I always seemed to be having fun and never being short of time. I always had the time to offer a cup of tea to anyone who ever walked into my office and listen to him as long as he wanted to speak. I nurtured no ambition to be a General. My only ambition was to do the job at hand, better than anyone else and be worthy of emulation by my command. And that was fun. My motivation came from the uniform that I wore and never from the hope of ranks. That never waned, promotion or no promotion. In doing so, I retained the control of my happiness, looked everyone in the eye and my ability to respond to my circumstances as I wished.

In retrospect, one of my shortcomings was that I never had fear of the authority, senior commanders, or anything else, right from school. Actually, there was disdain against unworthy superiors and that caused some trouble. I was an eternal optimist, who strongly believed in positive outcomes. I had no belief in God, per se. But I had no problems with what others believed in.

I rarely went to the temple or any other religious place. When I was the commanding officer, I made the weekly mandir parade a voluntary affair, much to the chagrin of the unit religious teacher. It was based on one belief that if a soldier was to live by the military values and all that this profession and the uniform stood for, there was no need of any religion. These values placed far more stringent codes of conduct on a soldier than any religion did. And motivation of our commands must be based on military values, for soldiering is our religion.

I miss the spit and polish and the bugles and the drums. I miss the resounding salute of 'Jai Hind, saab'. I miss the clatter of gunfire and booming sounds of gun and tank fire.

I miss the deafening roar of fighter jets flying overhead. I miss the noise and commotion of training with these men. I miss those anxious moments when the phone rang in Lolab. I miss the environment of trust and the smiles of the men. I miss those lovable soldiers who made this journey so wonderful.

I miss the glint of the medals and shining brass buttons. I miss the barakhanas and sharing a peg of rum with the men. I miss the langer dal soup in a big steel glass. I miss rubbing shoulders with the men in games and training. I miss those yelling dedicated drill sergeants, who took upon themselves to shape the future of the Indian Army. I miss the silence and the ambience of unit lines and military cantonments. I miss the roar of military trucks whizzing past in long convoys. I miss those wonderful buddies who would wait for you till late night and were prepared to risk their lives for you.

Above all, I miss the uniform and what it stood for.

I know, I have been critical of many things in the book. It is because I love my Army and as righteous soldiers and the last bastion of the nation, I expect very high standards from officers, as I demanded from myself and followed. There simply cannot be 'chalta hai' in the Army. The expectations from the countrymen are high, so are the stakes. The higher a person goes, the higher is the expectation from him for competence, personal example and integrity. Sadly, that is the confidence that the higher hierarchy has not been able to instil in the subordinates. The nation will call upon the Army once in a long while—we can never fail—not once. It won't happen unless we are ready every day. There is simply no comparison with any other service—we have our own standards and codes to live by.

I lived the path which was tough, but loved living that way. If I had lived the pedestrian way, I would be just another man who walked the beaten trail, unnoticed, making no impact on a man or a dog. I stood by what was right, no matter who the person was or the consequences. I stood by my subordinates when the others did an about-turn, and served selflessly.

In retrospect, I tried to change the system alone, right from day one. It does not happen. I may have cornered a few Brigadiers and Generals; more of the same came along the way. What one needs to do is to set a compelling example for those around you so that they emulate your values and pass them along the way. This will ensure that in due course, a small trickle becomes a deluge and the culture changes. Trying to do it alone, as I tried, will consume you making little difference to the organisation. I do believe that I set an example that many see worth emulating. And that will make some difference, however small.

We must remember, the soldier is the army. He remains the most important element of the army. He is not here for us, we are here for him. We do not have to convince our superiors; every officer has to convince his men, that he is worthy of their command and trust. It is only then that the army will be a professional army.

No institution has ever earned the trust of our countrymen as has the Army. It is our obligation to add to that reservoir of trust and faith, through character, integrity and commitment to the military values, drop by drop. In this, there are no fixed deposits.

A belief that I nurtured as a young boy in school that one day I would write a book about myself has fructified. I have written this with complete belief and truth. I have sacrificed humility for this narrative to be authentic.

I know, it has not been easy. It needs guts for cadet in the NDA to tell the squadron commander that he has been sleeping and not me, a young captain to question a commander on his leadership, or write to the Chief that his report is wrong. I did that repeatedly when I was convinced, when others sat dumb.

I did that for one simple reason; I just could not look the other way. For me, it was a compulsive response to the violation of military values by any officer, junior or senior. I was made that way.

For me, my present rank was my last rank. I do not regret doing anything in service. On the whole, I enjoy a satisfying reputation as a human being and as a soldier of substance. It was the result of the way I chose to live and serve—fearlessly, honestly and selflessly.

Will I do it all over again? Every time I am born, I would love to be a soldier of the Indian Army, for one simple reason;

> *"Where else will you find such magnificent company of men and women whose life is based on trust, faith and selfless service?"*

As for my lack of humility in this book—I see no need to apologise. You are not bragging if you have done it.

* * * * *

INDEX